Historic Textiles, Papers, and Polymers in Museums

ACS SYMPOSIUM SERIES **779**

Historic Textiles, Papers, and Polymers in Museums

Jeanette M. Cardamone, EDITOR
*Agricultural Research Service, U.S. Department of
Agriculture*

Mary T. Baker, EDITOR
Egypt MVE

American Chemical Society, Washington, DC

Library of Congress Cataloging-in-Publication Data

Historic textiles, papers, and polymers in museums / Jeanette M. Cardamone, editor, Mary T. Baker, editor.

   p. cm.—(ACS symposium series ; 779)
   ISBN 0–8412–3652–6 (alk. paper)
   1. Textile fabrics—Conservation and restoration—Congresses. 2. Paper—Preservation—Congresses. 3. Polymers—Conservation and restoration—Congresses. I. Cardamone, Jeanette M., 1959– II. Baker, Mary T. III. Series.

TS1449.H58   2000
702'.8'8—dc21
                             00–058618

The paper used in this publication meets the minimum requirements of American National Standard for Information Sciences—Permanence of Paper for Printed Library Materials, ANSI Z39.48–1984.

PRINTED IN THE UNITED STATES OF AMERICA

Foreword

The ACS Symposium Series was first published in 1974 to provide a mechanism for publishing symposia quickly in book form. The purpose of the series is to publish timely, comprehensive books developed from ACS sponsored symposia based on current scientific research. Occasionally, books are developed from symposia sponsored by other organizations when the topic is of keen interest to the chemistry audience.

Before agreeing to publish a book, the proposed table of contents is reviewed for appropriate and comprehensive coverage and for interest to the audience. Some papers may be excluded in order to better focus the book; others may be added to provide comprehensiveness. When appropriate, overview or introductory chapters are added. Drafts of chapters are peer-reviewed prior to final acceptance or rejection, and manuscripts are prepared in camera-ready format.

As a rule, only original research papers and original review papers are included in the volumes. Verbatim reproductions of previously published papers are not accepted.

ACS Books Department

Contents

Polymers in Museums

Preface

This volume continues the fine tradition of the American Chemical Society (ACS) of publishing papers from the symposia of Historic Textiles and Paper symposia sponsored by the Cellulose, Paper, and Textile Division of the ACS. Three Historic Textiles and Paper symposia have been published as ACS Symposium Series and two as ACS Advances in Chemistry Series. Two of the books were edited by Howard Needles and Haig Zeronian, retired Professors of Textiles of the University of California, at Davis. Needles and Haig have made significant contributions to the conservation science literature. In keeping with this tradition, this book is a tribute to the ACS for its promotion of the science base of fundamental investigations for historic works of art. This volume presents reports from two symposia: *Historic Textiles and Paper* and *Polymers in Museums* that were presented at the national ACS meeting in Boston in 1998.

Within the symposia, investigators have examined unique characteristics and analytical methods that have lead to proposed solutions for stabilizing aged protein fibers (wool and silk) and cellulosic fibers (paper and wood) as well as synthetic polymers. These investigators are practicing conservators, conservation scientists, and academic researchers. Their works add to the rich resource of conservation research that informs broad audiences of the unique challenges involved in investigating objects of historic significance.

This ASC volume on Historic Paper and Textiles is an important reference for the application of scientific methods to examine artifacts and art objects and for the assessment of their condition. It can be used for formulating decisions concerning documentation, prescribed care, exhibition, and storage. It covers the aging of natural and synthetic polymers such as wool, cotton, silk, paper, wood, and polymers used in sculptures and photographics films as well as in other media. The aging properties of wool fibers and a comprehensive review of the aging, degradation, and conservation of historic materials made from cellulosic fibers will prove an invaluable resource for conservators of historic objects, art historians, academicians, and curators of museum collections who are responsible for the conservation and restoration of museum artifacts.

This book examines the influence of mordants and natural dyes on the color fading and the degradation of cotton and silk fabrics and their implications for determining conditions for exhibition and display. Important analytical methods of high-performance size exclusion chromatography, viscometry, and gel electrophoresis are described as methods for assessing damage in silk textiles. The technique of microchemical manipulation, light microscopy, and FTIR spectroscopy are described for the analysis of historic cotton fibers including those from a marine environment. Other spectroscopic investigations include the degradation of vulcanized natural rubber museum artifacts. Laser surface profilometry is introduced as a novel technique for the examination of polymer surfaces. Also included is a comparison of natural and artificial aging effects of cellulose acetate artifacts as followed by FTIR spectroscopy and ion chromatography, and the applications of light microscopy and FTIR spectroscopy to distinguish between oil and acrylic painting media. This book will serve as a valuable tutorial and resource for the understanding of the properties of new and aged works of art across a broad spectrum of media.

Aged historic objects present special problems and require unique solutions. Addressing the degradation of an aged object can require incorporating conventional and new technologies. The ACS is among the few publishers who offer basic and practical information on the application of science to the conservation of works of art. The time is ripe for such a publication to address the U.S. Government's Third Millennium Initiative that focuses on the preservation of historic objects.

JEANETTE M. CARDAMONE
Eastern Regional Research Center
Agricultural Research Service
U.S. Department of Agriculture
600 East Mermaid Lane
Wyndmoor, PA 19038

MARY T. BAKER
Chemonics International
Egypt MVE
1133 20th Street, N.W.
Washington, DC 20036

Textiles and Papers
in Museums

Chapter 1

Historic Textiles and Paper

Jeanette M. Cardamone

Eastern Regional Research Center, Agriculture Research Service, U.S. Department of Agriculture, 600 East Mermaid Lane, Wyndmoor, PA 19038

The care of historic textiles and paper materials presents unique challenges for those working in conservation and restoration. The challenges are to maintain or reinforce the structural integrity and aesthetic appreciation of these objects while preserving authenticity through appropriate selection and application of materials and methods. The uniqueness of each object presents specific challenges for formulating decisions that not only require experience gained from similar pieces but also specific knowledge of materials and a firm understanding of the chemical and mechanical behaviors of all components of the object. Thus experience and knowledge become the underlying and fundamental basis for decisions regarding care.

Historic textiles provide priceless records of the history of mankind. They are doubly significant because of their inherent documentary value and that of the objects produced from them. As depictions of the art and ethics of mankind with cultural, social, and political relevancy, historic textiles continue to inspire multidisciplinary studies. The marvel of textiles is in their construction from individual fibers that are assembled for processing through numerous steps and stages of yarn to fabric formation. Each fabric is a reflection of the aesthetic sense of the fabricator and the knowledge of how the sequence of processing steps such as scouring, bleaching, dyeing, and finishing can modify appearance. As one of the oldest skills, textile fabrication can provide a record of technological developments throughout the ages. Often the historic textile is composed of several different kinds of materials and some contain secondary materials, resulting from adding new to old, thereby conflicting the design and hand of the original maker.

This complicated set of problems presents severe challenges for the conservator who restricts treatment to preserving the material and human evidence of the object. Alternatively, the restorer will strive to apply treatment to return the textile to some semblance of its original appearance, structural integrity, and aesthetic quality. The predominant theme for the care of historic textiles is to preserve the authenticity of the object by appropriate care beginning with the selection of materials and methods of treatment.

In the case of a textile of great historic significance that has survived in a fragmented state, decisions may be made to either replace the missing areas with sympathetic materials that are recognizably newer additions to enable the aesthetic appreciation of the piece, or to leave it in its original state. Fragmented textiles can be restored to the semblance of the original object by consolidating the remaining portions between transparent netting. For example, the author, as part of a restoration project at the Textile Conservation Centre, formerly at Hampton Court Palace, Surrey, England, partially restored an 18th century swallow-tail silk Guidon to its full size and shape with silk fabric of similar weight and construction that was dyed to the same shade as the original, before sandwiching it between silk crepeline that was dyed to the same color. When an historic textile is so severely degraded and friable that it cannot be exhibited, a decision may be made to consolidate it by casting a polymeric film to fill in missing areas in order to restore its original form. This method would be least desirable for it compromises a primary principle that restoration methods be reversible.

Sometimes the historical significance of a piece extends to the deposited residues. The Shroud of Turin, thought to be the burial cloth of a crucified man (affirmed by some to be Jesus Christ), exhibits marks resembling blood residues that are positive renderings on a linen fabric bearing the photographically negative body image. The juxtaposition of these positive and negative images is of great importance for scholars who investigate the formation of the image for it suggests that the blood-like marks and body image were formed by different mechanisms. In another case, a heavily blood-encrusted military uniform might lose its fabric stability should it be subjected to cleaning before conservation or restoration.

Woven tapestries present a unique challenge. In a "Great Tapestry" where there is an interlacing of highly twisted wool warp yarns with weft yarns of low twist, subjecting the fabric to wet cleaning will cause fiber swelling with differential yarn untwisting because the yarns will take up a greater surface area. Pockets and wrinkles will form throughout the fabric surface and they will act as light traps. The result is that the once youthful, mythological figures depicted in the tapestry will appear to be prematurely aged and this will diminish the aesthetic appeal of the piece. Thus, not only is it important to document the state of deterioration of a historic textile, but at the onset of investigation, it is extremely important to document thoroughly all textile components of the object and the object's past history. Complete understanding of all implications is critical for the appropriate care of a historic object.

The fading characteristics of natural dyes, some mordanted with natural salts, is another consideration because they were used prevalently in historic tapestries. Generally, natural dyes exhibit poor light fastness. In some cases, the once green foliage and grass of old tapestries appear blue because of fading of the yellow dye component that was used in combination with blue to form the green color. The stability of color can be dependent on dye structure, the presence of a mordanting salt, dyeing process, fiber type, quality and intensity of incident light, environmental exposure, relative humidity, and the ambient temperature.

Natural fibers are proteinaceous (wool and silk) or cellulosic (cotton, paper, flax, hemp, and jute) in chemical composition. Within the classes of protein and cellulosic fibers there are not only differences in chemical reactivity but also differences in fiber morphology. When the fibers of these classes are compared, they exhibit diametrical chemical and mechanical properties. Different handling procedures and sets of conditions for treatments that are time-, temperature-, and concentration-dependent are required. Within certain time, temperature, and concentration ranges, protein and cellulosic fibers respond differently: wool fibers can be degraded by base and not by acid whereas cellulosic fibers can be degraded by acid but not by base. Chemical and morphological differences dictate how applied chemical, physical and mechanical forces will affect a historic textile. According to morphology differences, wool's mechanical profile is "soft" and extensible (due to a molecular coil configuration), whereas silk's profile is less pliable because its molecules are laid down in sheet-like form. Thus wool is more resilient and less prone to wrinkling. Mechanically, the cellulosics are relatively hard and brittle with little resiliency. They tend to deform permanently under mechanical stress so that wrinkles are retained. These factors influence every aspect of care including accession, documentation, conservation, restoration, exhibition, display, and storage.

The degradation of wool is addressed in a chapter by Weatherall who postulates a mechanism for chemical changes resulting from light exposure that will aid in understanding the degraded state of a woolen artifact. In another chapter, Cardamone describes the fiber morphologies and characteristics of cotton and flax fibers and their modes of degradation. These chapters reinforce that the useful lifetime as a viable historic textile is limited by the stability of the basic chemical and morphological structure even though the textile object exhibits a somewhat stable fabric construction.

Chen and Jakes describe the aging properties of marine cottons by documenting changes in crystallinity indices using FTIR microspectroscopy and microchemical reactions for analyzing chemical and physical changes and internal structural differences in dyed and undyed marine cottons after treatments with swelling agents. The degradation and color fading of prototype cotton fabrics dyed with natural dyes associated with metal salts were examined by Kohara, et al., and of prototype silk fabrics by Yatagai, et al. The authors describe the changes in mechanical properties and the influence of natural dyes and mordants on the aging of cotton and silk as monitored by inductively coupled plasma atomic emission spectroscopy (ICP-AES) and electron spin resonance (ESR). A study by Tse describes the differences among three analytical methods; high performance size exclusion chromatography (HPSEC), viscosity, and sodium dodecylsulfate polyacrylamide gel electrophoresis (SDS-PAGE) for evaluating the degradation of silk after artificial light aging.

The topical effects of aging are evident as discoloration that is accelerated by moisture and heat. Historic paper material is aged cellulose with far greater surface area than a cellulosic textile of comparable weight so the effects of aging can be more dramatic. The aging of paper and its permanency first became a concern when books of newer vintage showed signs of degradation while books of over 100 years remained in good condition. A possible cause was residual acidity from sizing and alum, introduced by newer paper making processes. Correlations were drawn between acidity and extent of degradation. Causes and remedies were studied in terms of reasonable life expectancies by carrying out accelerated aging with extrapolations to current conditions. The chapter by Erhardt et al. compares the effects of artificial and natural (dark) aging on mechanical properties by following the production of glucose, the ultimate product of cellulose degradation, as detected by gas chromatography. From this study, recommendations were derived for optimal conditions that favor stability.

The preservation of materials derived from natural fibers was a solely primary concern until the turn of the 20th century when man-made fibers made inroads into the manufacture of apparel and home furnishings. Man-made cellulosics: viscose rayon and cellulose acetate from the 1920's and 30's and the synthetic fibers such as nylon and polyester from the late 1930's to '50's offered materials with new and different fiber properties and modifications that originated from selective processing conditions, dyes, dyeing procedures, and finish applications. At times man-mades and synthetics are selected as conservation materials because they overcome some limitations of natural fibers. However, specific considerations must be for the selection of support fabrics used to stabilize degraded textiles. For example, natural fibers with high moisture regain (wool, 13.6% - 16.0% and cotton, 8.5%) are more likely to support microbiological growth and insect contamination whereas nylon with low regain (4.0%) and polyester with only slight moisture regain (0.4%) can resist microbial growth but will be prone to attract more dust through static accumulation. (1) Other environmental factors such as temperature and humidity, water content, pH of exposure, cleanliness, and air pollution influence the extent of degradation of natural and synthetic fibers and their susceptibilities to microbiological decay and insect attack. Biological damage can be assessed by visual inspection aided with microscopical methods, and chemical, microbiological, and enzymatic instrumental methods of analyses.

As organic materials, historic textiles and paper objects are in a constant state of progressive decay by exposure to ambient oxygen and light exposure, acids and bases, oxidizing and reducing agents, and bacteria or fungi which utilize enzymes to hydrolyze polymer chains for main chain scission or in the case of wool, the digestion of wool at the sites of covalent disulfide linkages that join molecular chains together. The overall deterioration of cellulosic materials can be followed by examining the effects of hydrolysis and ensuing acidity and by analyzing for hydroxyl-to-carbonyl group conversion where the number of reducing carbonyl groups is indicative of main chain scission resulting in shorter molecules and reduced fiber strength. The overall deterioration of wool can be followed by the oxidation of disulfide bonds to sulfoxide groups, and the hydrolysis of main-chain peptide bonds that causes intermolecular bond cleavage resulting in strength loss. The amino acids of the peptide chain that are most susceptible are histidine, tryptophan, tyrosine, methionine, and the sulfur-containing amino acids, cystine and cysteine. Silk and wool as polyamide structures also can oxidize to form carbonyl and carboxylate groups. The degradation of nylon can be followed by the formation of amine and carboxyl groups and for polyester, by the formation of carboxyl end-groups. Analytical spectroscopic methods of analysis are invaluable for recording the development of these functional groups to indicate the extent of aging. At the structural level, optical microscopy and scanning electron micrographs of aged fibers provide dramatic evidence of splits and wear that are clues to the causes of deterioration.

Dyes and chemical additives may promote aging or provide photoprotection. Vat dyes can cause photosensitization and phototendering of cellulosic textiles. (2) In undyed textiles, yellowing is regarded as evidence of degradation and becomes manifested in the aging of cellulose through photooxidation and accompanying dehydration. These processes cause the formation of conjugated ethylenic carbonyl groups that absorb visible light in the blue region and reflect the yellow color. The yellowing of wool has been attributed to oxidation of specific amino acids, tyrosine and tryptophan that can be accompanied with formation of carbonyl groups from any residual waxes and lipids that can be present in residual suint (natural wool grease). Silk can yellow by the oxidation of the amino acid, tryptophan. Instrumental methods of analyses can be used to document these changes in molecular structure. Alternatively, classical, noninstrumental, wet-chemical methods of analyses are available for conservation research. For example, damage to cellulose can be assessed by carboxyl group determination, copper number, fluidity, alkali swelling, solubility in specific solvents, and fiber strength measurements. Damage to wool can be assessed by alkali and urea-bisulfate solubility. The ninhydrin (triketo hydridene hydrate) reaction can be used to assess damage to wool by qualitative colorimetric assay of amino groups in wool and silk protein substrates, and in nylon. The reader is referred to the book, *Analytical Methods for a Textile Laboratory*, 3rd Edition, J.W. Weaver, Ed., American Association of Textile Chemists and Colorists, Research Triangle Park, NC, 1984 for more complete information.

Although the degradation process is ongoing, it can be slowed, retarded, or even arrested depending upon physical and environmental conditions and by applying compounds designed to stabilize it. Generally, the degradation of mechanical and chemical properties is a manifestation of underlying chemical changes. Often with the application of science to works of art, the degradation process is simulated and studied methodologically so that cause and effect can be ascertained. In a simplistic view, with broad sweeping assumptions that follow precedents for artificial aging used in commercial settings, the overall aging process can be simulated to estimate the lifetime of the material when exposed to a predictable set of conditions. Useful information can be derived to predict the aftereffects of treatments designed to impede the aging process. An alternative approach is through the application of analytical methods to characterize the present state of the object and to follow the progressive deterioration of the object over time. The application of science to document, investigate and preserve historic textile and paper materials is crucial as these diverse objects of posterity continue to challenge the museum curator, the conservator, and conservation scientist to a full understanding of the properties of these materials and the forces that alter them.

References:

1. Joseph, M.L., *Introductory Textile Science*, New York: Holt, Reinhardt, & Winston, 1986, p. 22.

2. Trotman, E.R., *Dyeing and Chemical Technology of Textile Fibres*, 6th Ed., New York: John Wiley and Sons, 1984, p. 426.

Chapter 2

The Aging, Degradation, and Conservation of Historic Materials Made from Cellulosic Fibers

Jeanette M. Cardamone

Eastern Regional Research Center, Agricultural Research Service, U.S. Department of Agriculture, 600 East Mermaid Lane, Wyndmoor, PA 19038

Cotton and flax fibers, although different in fiber morphology, have somewhat similar physical and mechanical properties. Their aging properties and degradation profiles can be studied through the chemistry of cellulose. The factors that degrade these cellulosic substrates can be environmental, resulting from oxidation, hydrolysis, thermolysis or pyrolysis, gaseous exposure, microbial attack, and physical stress that can cause disruptions in macrofibrillar structure to impact on mechanical stress and strain. Various modes of attack that advance the progression of aging in cellulosic fibers are examined for relevance to changes in fiber morphology and chemical behavior. Reference is drawn to aging studies to elucidate degradation mechanisms and to derive kinetic data that can have value for predicting the future effects of treatments.

Cotton, an old-world fiber, may have grown in Egypt in 12,000 B.C. and India in 3000 B.C. India is generally recognized as the center of cotton production since 1500 B.C. until the early 16th century (1,2). Cotton was indigenous to North and South America, Asia and Africa, and Peru. In the United States, cotton cultivation goes back to 500 B.C. in areas known today as Utah, Texas, and Arizona based on fragments that have been C[14] dated. The ancient Anasazi, predecessors to the Southwestern Pueblo Indians, cultivated cotton in the Southwest. The first recorded date for the planting of cotton in the US was 1536. Soon colonies, particularly Virginia and in North and South Carolina clothed one fifth of the population in cotton. A series of inventions including mechanical means to separate the cotton plant's fiber from seed and the Eli Whitney cotton gin increased the cash value of the cotton crop. Today cotton is produced in the beltwide region of the US that includes the Far West, Southwest, Mid-South, and Southeastern states where the climate is temperate to hot with adequate rainfall or irrigation.

Natural cellulosic fibers were used to fabricate lawn, batiste, muslin, gauze, chintz, and calico cottons and linens throughout the 18th century Indienne craze that captivated the settlers of colonial America. With the invention of the power-driven knitting machine in 1863, cotton gradually became a replacement for woolen underwear and was featured in Sears advertisements for bathing suits, shirts, and dresses. Linen, the fabric from flax fibers, maintained the aura of "cleanliness" from the time of the Renaissance when it was preferred for borders and collars that were fresh and clear to emphasize the neckline of a person above the industrial ranks (3). From this time to the present, linen continued to be preferred for household use, underclothes, table linens and handkerchiefs.

Cotton fiber is the seed hair of the plant species, "Gossypium." These seed hairs grow to maturity within ripening cotton bolls that remain on the plant after the flowers fall off. The seed hairs are separated from seed and are composed mainly of cellulose with a thin coating of wax that is mainly removed in fabric finishing so that cotton absorbs moisture readily. Cotton fibers vary from 16 to 20 micrometers in diameter and are classified within staple length ranges of short (less than 1 inch), medium (1" to 1 1/8"), long (1 3/8" and longer). Other important properties are strength, fineness, color, maturity, uniformity, and luster. Cotton's lack of resiliency was improved by one of the greatest technological advances in fabric finishing: resin-finishing to impart "easy-care" performance (4-6).

Fiber Morphology

The microscopic images of cotton fiber cross-sections reveal a characteristic kidney-bean shape and longitudinal views show a twisted or convoluted structure. Reportedly, long-staple cotton has about 300 convolutions per inch and short-staple cotton, less than 200 (7). The fiber is composed of layers of cuticle, primary and secondary walls, and a lumen (the hollow central core). The cuticle is the outside skin that provides protection against chemicals and degrading agents. It can be removed by kier boiling and bleaching so that cotton becomes more absorbent and more receptive to dyes, yet more readily affected attack by solutions. Subsequent launderings remove cuticle residues and leave the fiber vulnerable to increasing degradation through use, care, wear, and refurbishing. The primary cell wall contains fibrillar cellulose, 200 nm thick, where fibrils are found within a spiral configuration. The characteristic spiral of 70 degrees to the fiber axis provides strength to the primary cell wall and ultimately to the fiber as it forms. Within the secondary cell wall are concentric layers of spiraling cellulosic fibrils, 10 nm thick, aligned 20 to 30 degrees to the fiber axis. Thus strength is imparted as well as stability. Most cellulose content is concentrated in the secondary wall. The lumen or hollow canal forms at a certain growth stage when aqueous proteins, sugars, and minerals evaporate to cause a pressure differential within the fiber so that it collapses to the form of the characteristic kidney bean shape used to identify cotton fibers microscopically. Overall, cotton fibers contain 90–95% cellulose, 1-2% protein-aceous materials, 0.3–1% waxes, and small amounts of organic acids and ash producing inorganic substances.

Other cellulosics: flax, ramie, jute, and hemp, are bast fibers originating in the long stems of these plants. Flax belongs to the genus Linum of the family Linaceae. This group includes numerous species including *Linum usitatissimum L.* The species was known in Egypt, Phoenicia, Mesopotamia, Greece, and Rome and is cultivated today. Fragments of 3000 year-old mummy linens from Egyptian tombs represent some of the oldest archeological textiles preserved in museum collections today. Cultivated and wild flax species, including the most ancient, *Linum Augustifolium*, know to have grown in Palestine, have been used throughout antiquity. Remarkably, Egyptian linen fabrics were known to be as fine as a web, with yarns that could be set to 540 yarns per inch (8). It was reported that the Hebrews were skilled linen weavers and that they imported linens from Egypt for use in religious celebrations to signify purity. Fine twilled linens were used as tabernacle closures (9). The fineness of these linens is a tribute to the handweavers of that time, practicing highly advanced textile technology.

Flax fibers grow as single cells in the stem of the plant. They range in length from 12 to 36 inches and reside between the bark and inner woody tissue of the plant. In the cross-sectional view of the stem, flax fibers have a polygon shape. These polygons grow together as bundles of fine fibers referred to as "ultimate cells", each as fine as 10-20 microns. Three to six ultimate fiber cells can comprise a cross-sectional fiber bundle having a diameter of 30 to 120 micrometers. When a bundle of flax fibers is reduced to only a few ultimate cells through bleaching, appreciable strength is lost. Unlike the spiral configuration of unicellular cotton, flax fiber bundles are aligned side-by-side along the fiber axis so that when tension is applied from either end, the load is borne axially and is uniformly distributed, resulting in a higher breaking load. Processing flax to ultimate cells for the spinning of fine yarns involves a retting or microbial process to degrade the bark and subsequent steps to remove it from the fiber bundles. Bleaching to the white stage could cause a 27% loss in weight and a 28% loss in breaking strength (10). By this process, stiff, straw-colored flax becomes supple and white. The connecting matrix for flax fiber bundles is a combination of lignin and associated natural materials such as hemicellulose, pectins, water-soluble compounds, fats, and waxes. Although lignin is present at 2.8% by weight in unretted and 2.2 % in retted flax, it provides a stress-transferring, amorphous matrix when stiff flax fibers bear a load. Lignin is mainly confined to the middle lamella and outer walls of a few of the ultimate fibers. Its presence is responsible for the characteristic straw-color of flax. Lignin enhances the resistance of the fiber to bacterial attack. It alters the properties of the fiber, and its aromatic structure is an ultraviolet light absorber so that it can absorb radiant energy of short wavelengths that may be sufficient to break chemical bonds resulting in strength loss. Although the presence of lignin has positive and negative implications, the enjoyment of white linen requires that lignin be either bleached or removed. The cellulose content of unretted flax fibers is 62.8% and of retted is 71.2% by weight. Hemicellulose content of unretted flax is 17.1% and is18.5% in retted. Lignin content of unretted flax is 2.8% and is 2.2% in retted. Other associated substances including fats, waxes, colorants, and impurities to make up 8.1% by weight (11).

The fiber morphological differences between cotton and flax are shown in Figures 1 and 2 below.

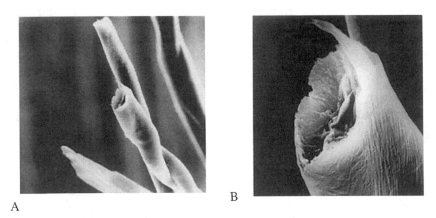

*Figure 1a,b. Scanning electron micrographs of new cotton fibers showing:
a. Spiraling fibrillar structural features, convoluted detail and hollow central lumen,
2,000x. b. 10,400x magnification.*

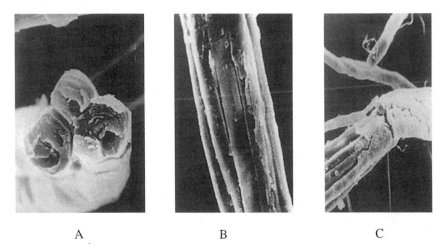

*Figure 2a,b,c. Scanning electron photomicrographs of flax fibers: a. One new flax
fiber bundle containing three ultimate flax cells, 1,000x. b. A new flax fiber
consisting of a bundle of several ultimate cells encrusted within the
hemicellulose/lignin cementing matrix, 740x. c. A flax fiber bundle subjected to
190C showing cracking of the outer encrusting medium, 840x.*

Fiber Properties

A comparison of the physical and mechanical properties of cotton and flax
celluloses is shown in Table 1.

Table 1. Physical and Mechanical Property Differences of Cotton and Flax Cellulose (12)

Fiber Type	Tenacity Grams/denier[a]		% Elongation[b] at Break		Moisture Regain	% Elastic Recovery[***]
	Dry	Wet	Standard[*]	Wet	At 65% RH, %[**]	
Cotton	3.5-4.0	4.5-5.0	3-7	9.5	7-11	75
Flax	3.5-5.0	6.5	2.0	2.2	12	65

	Work of Rupture (13) (g-wt/tex)	Initial Modulus (13) (g-wt/tex)
Cotton	1.52	740
Flax	0.82	1830

[a]g/den = (N/tex) × (11.33).
[b]Stretch at rupture.
[*]Standard conditions of 65% relative humidity and 70°F
[**]Expressed as a percentage of the moisture-free weight at 70°F and 65% relative humidity.
[***]Recovery from 3% stretch.

Both cellulose and flax exhibit a high modulus with no yield point. Under applied stress, these fibers resist strain linearly up to the breaking point. When work of rupture and initial moduli are compared, flax, being harder and stiffer, is more susceptible to brittle fracture. From Table 1, flax fibers can only be extended 2% of original length before the breaking point and recovery after extension is poorer than that of cotton. The recovery properties after extension of both cotton and flax are poor. A release of the stress at any point below the breaking extension leaves the fibers in a state of permanent deformation. The wrinkling and buckling of flax and cotton fabrics that result from permanent deformation can be somewhat alleviated by placing the textile in a humid environment to allow the fibers' inherently high regain to ease and relax creases. Furthermore, frequent cycling of applied stresses as occurs in handling, conditions the fabric to an increased state of embrittlement because the extension at break will begin to decrease even more. Because cotton and flax are brittle, little work is required for rupture. These limitations can be compensated by careful selection of yarn and fabric characteristics. Fine yarns and open fabrics with low fabric counts (ends per inch x picks per inch) are most susceptible to this type of failure.

Cellulose Structural Change

Both cotton and flax are degraded by acid. The severity of degradation is determined by exposure time, temperature of application, and reagent concentration. These fibers are susceptible to chemical attack by exposure to bases, oxidizing and reducing agents, microbiological attack by bacteria or fungi which utilize enzymes to hydrolyze polymer chains, physical agents such as ultraviolet radiation, and changes in crystalline state. The degradation effects can be negligible or so severe that the textile loses its strength and is no longer viable for its intended end-use. Natural or physical degradation (aging) can proceed homogeneously throughout the substrate. It can proceed heterogeneously from the outside surface through to the interior, generally with loss of structural integrity to the amorphous regions, leaving a highly crystalline substrate behind that now exhibits different physical and mechanical properties. In this case, the fibers become even harder, stiffer, with even less ability to recover from physical force. The percentage of crystalline material in native cotton is approximately 60% (14). Interestingly, aging effects in natural fibers can be somewhat reversed by wetting them (albeit other considerations must be made before wetting). Wetting lowers a fiber's glass transition temperature, T_g. The term applies to a synthetic polymers' softening temperature below which the polymer is hard and brittle and above which there is molecular flow or the softening that is necessary to set a particular configuration. Because natural polymers do not flow, the term, T_g, is generally used simply to draw an analogy.

At the molecular level, cotton and flax exhibit the same chemical reactivity of cellulose, although their different morphologies in part determine their relative susceptibilities to attack by degrading agents. Cellulose, a homopolysaccharide, is a natural, long chain, linear polymer that is formed from the condensation of two glucose ($C_6H_{12}O_6$) molecules. Water is eliminated to form the basic repeat unit, cellobiose (anhydroglucose or beta-D-glucopyranose units). These units are linked together by 1-4-glycosidic bonds) that are combined approximately 10,000 times to form the polymeric structure shown in Figure 3.

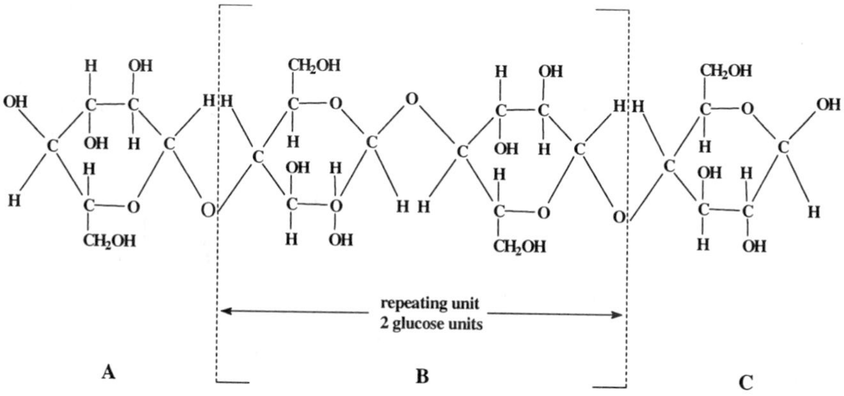

Figure 3. The partial molecular structure of cellulose ($C_6H_{10}O_5$) in the 1,4-beta-D-glucopyranose form showing the repeating unit (B) composed of two anhydroglucose units. Portions (A) and (C) are terminal units of the molecule where "C" can form an aldehydic reducing endgroup.

Note that the cellulose molecule is a polyhydric compound (many hydroxyl or -OH groups). There are three hydroxyl or methylol groups for every anhydroglucose unit. Consequently there is a high degree of hydrogen bonding between the long and linear polymer chains that are stacked in close association through Van der Waals physical forces of attraction and the geometry of the short carbon-hydrogen bond distances. The symmetrical arrangement of anhydroglucose linear chains causes the formation of a closely packed polymer structure. During fiber development, bundles of cellulose molecules are either associated in a highly ordered state parallel to each other to form crystalline regions or in disarray to form amorphous regions. Microfibils associate to form fibrils, and then to macrofibrillar bundles of higher order that provide the structure for insoluble fiber with high tensile strength.

Cellulose Degradation

The hydrolytic terminal end at C in Figure 3 is a "reducing end" for aldehydic (HC=O) reactivity. The reactivity of cotton and flax is in part predetermined by the general chemical reactions of the hydroxyl group that undergoes hydrogen bonding, etherification, esterification, oxidation, hydrolysis, and pyrolysis. Of particular interest to conservators are thermal and photochemical reactions as well as biological activity. All result in a decrease in molecular weight through depolymerization or scission of polymer chains, possibly with the inclusion of oxygen so that the oxidized species exhibits lower strength, elasticity, and toughness. When degradation pathways result in a higher degree of polymerization, the textile substrate suffers further embrittlement.

The reactivity of the hydroxyl groups is limited by the fiber's accessibility (the relative ease by which these groups can be reached by reactant). Accessibility is determined by amorphous content. Amorphous regions are more readily attacked by foreign agents than are crystalline regions. Drastic drying conditions with the loss of water (H-O-H) cause extensive interchain hydrogen bonding thus reducing the accessibility of the hydroxyl groups. The yellowness that is commonly noticed in aged cellulosic textiles may or may not impact on strength unless degradation has occurred through main chain scission. In the case of yellowing, aged cellulose's reduced capacity for hydrogen bonding, coupled with incorporation of ambient oxygen, can promote formation of conjugated ethylenic (-C=C-) / aldehydic (-H-C=O) chromophores. These chromophores absorb in the blue region of the visible spectrum, thereby conveying the visual perception of yellowness.

There is long-standing concern about discoloration that can result from storing cellulosic materials for a great length of time. In one case, a bale of cotton stored for fifteen years, had increased in yellowness yet it was still strong enough to spin into yarn, although strength loss had not been determined for lack of the opportunity to compare current fiber strength to original strength (15). Often, aged cellulosic textiles that have been wet-cleaned bear yellow, tan or brown water- stains and / or a brown line that forms at the liquid front. The effect has been attributed to acidic residues that can develop through oxidation *even in the absence of* light, or heat, atmospheric oxygen, iron, bacteria, and waxy materials (16).

In cellulosic solid substrates, the natural process of oxidation to form the oxygen-containing carbonyl and carboxyl acid groups known as "oxycelluloses", is less drastic than the industrial processes such as commercial bleaching where there can be a dramatic reduction in degree of polymerization. However, in the presence of an acid environment, cellulose textiles can undergo hydrolysis whereby the imbibed water causes decomposition. For example, the glycosidic bonds in cellulose are vulnerable to acid-catalyzed hydrolysis which can ultimately break it down to D-glucose, thereby causing damage to cellulose chains, microfibrillar bundles and dissolution of the fiber. The acidic degradation of cellulose is described in the review by Nevell (17). In fact, when cotton fabric was artificially degraded by radiation, acid, or heat, the acidic oxycelluloses that formed were attributed to the aldehyde groups that were originally formed having been converted to carboxyl groups (18).

Deacidification

Attempts to counteract the effects of acid hydrolysis and subsequent cellulose degradation have included the following: rinsing and washing to remove acidic residues and applying alkaline buffers (pH 8) such as calcium carbonate, barium hydroxide, calcium bicarbonate, magnesium bicarbonate, and methyl magnesium carbonate powder where the presence of a basic environment would cause no damage to associated materials. Kerr et al. (19) and Block (20) have shown by artificial aging that providing a cellulosic textile with an alkaline reserve can be beneficial in slowing oxidation.

However, Feller (21) cites work that reports high alkalinity has been known to result in the degradation of pulp products by way of an unzipping mechanism. This shortens the molecular chain length to the point of ultimate dissolution of the polymer chains.

Aged Cellulosic Textiles

The Shroud of Turin is a linen cloth, 14 feet long and 3 ½ feet wide, constructed as a three-to-one herringbone twill with approximately 39 end per inch × 26 picks per inch. Images that might be attributed to aged cellulose appear head to head, as though a body had been laid on its back at one end of the fabric, which was then drawn over the head to cover the front of the body. The images reside on the tops of the linen yarns to indicate that if cellulose degradation were the source of the image, the aging effect was topical. Speculation for the image formation has included the possibility of a sudden heat event, the "Resurrection" perhaps, that would have caused accelerated heat aging at the points of body contact by anointing substances on the fabric. In laboratory simulations, it was found that baking modern linen in air at 150C for about four hours artificially aged the linen so that the discoloration replicated the background area that discolored from the passage of many years. The discoloration from heat aging was attributed to loss of water through the breaking of the interatomic bonds of cellulose to form $-C=C-$ and $-C=O$ functional groups and that "the addition of foreign substances that absorb the heat or light energy catalyzes or advances the rate of discoloration" (22).

In other work, Kleinert examined ancient linens from 2000 B.C. and 1500 B.C. by microscopy after their alkali swelling and found transverse dislocations, not found in modern linen. They were attributed to chemical attack and autooxidation. Although the crystallinity when compared to modern flax fibers was high, the strength was too low to measure. Water extracts were brown and acidic (pH 5). Infrared absorptions using potassium bromide pellets in the transmission mode after exhaustive water extraction, were found between 1650 and 1550 cm-1 with maxima at 1630 and 1610 cm-1. These principal bands were made up of separate absorptions at 1625, 1615, 1605 that were attributed to unsaturated carbon-carbon bonds, possibly conjugated double bonds in a cyclic system. A band at 1580 cm-1 could be attributed to the O-H bending of ionized acid groups. It was concluded that long-time natural aging shortened cellulose chains and created IR absorptions representing oxidative transformations of the native reducing end aldehyde groups to acid carboxyl groups. These results were duplicated in modern linen subjected to artificial heat aging and it was found that the basic mechanisms for degradation were similar, although the extent of aging was different. This suggests that accelerated aging of cellulose materials at elevated temperature, as commonly practiced as a test method, can indeed give useful information on their behavior during natural aging (23).

In another study, Berry and others investigated the aging properties of pre-Columbian cottons and compared them to the properties of contemporary cotton fabric subjected to accelerated aging by acid hydrolysis, heating, and irradiation with high-voltage electrons. Samples irradiated with 100-Mrad exposure for 90 to 120 hours exhibited greater acid concentration than when hydrolyzed by 40% sulfuric acid for 38 hours. In the modern aged fabric, oxidation developed by thermal degradation and by irradiation but not by acid hydrolysis and at a given strength level, the order of oxidation was the following: irradiation > thermal > hydrolysis. Natural and artificially aged fabrics showed a sharp drop in degree of polymerization with accompanying strength loss although there was not a direct correlation between strength loss and oxidized species nor crystallinity index. Thus there are different aging mechanisms. Overall, as the molecular weight decreased, microfibrils lost their structural integrity and the fiber failed by brittle fracture (24).

Cardamone and Brown (25) applied the Arrhenius relationship to the heat-activated rate of degradation of modern cotton fabrics at 100, 130, 160, and 190 °C in the presence of 50 lux tungsten illumination, in the dark, in air, and in nitrogen. The Energy of Activation (Ea) for the aging process was derived and used for room temperature estimations of the rates of degradation as measured by yarn strength loss. The Arrhenius model has long been established and used to estimate service life and to provide an estimation of failure distribution for a product (26). Empirically, kinetic data showed that the overall aging mechanism is constant over the 100 to 190 °C range, that is, that all the ensuing myriad chemical reactions can be subsumed under one unifying effect. Empirical data showed that first-order kinetics applied. From the kinetic data, half-life estimation of strength loss in the dark/air environment was 55.5% that in light/air. Thus there is an obvious advantage to storage in the dark. Applying the derived activation energy, Ea = 23,486 ± 1312 cal/mol, a 5 degree rise in temperature (20 °C to 25 °C) causes the rate of degradation to increase by 1.3. A 10 °C rise causes a rate increase of 1.8 and a 20 °C rise in temperature should result in a rate increase of 3.1. Increasing the temperature by 50 °C, possibly by photographing, would cause a rate increase of 16.8 (28). This methodology was applied for the nondestructive evaluation of degradation using reflectance absorption Fourier transform infrared (FTIR) spectroscopy (29). The chemical changes in cotton fabric, aged for 2,4,7,11,14, 23,and 31 hours in air and in nitrogen were recorded as carboxy group formation with the development and changes in the 1730 cm-1 absorption band. The rate of aging in nitrogen proceeded at a uniform rate over 31 hours whereas in air, a leveling off point was reached after 11 hours indicating the end of oxidative attack in the amorphous regions and the resistivity to chemical attack of the remaining highly ordered microcrystalline regions (30). It was reasoned that the increase in the 1730 cm-1 band resulted from the conversion of the reducing aldehydic end groups in cellulose to acidic carboxylate groups. It is recognized that other sources of these acid group formations include glycosyl bond cleavage and subsequent formation of shorter chains with end groups with hydroxyl and aldehyde that oxidize to carboxylate (31).

In another study, 1250 to 1300 A.D. cottons were analyzed by Photoacoustic FTIR and the absorption bands were compared to those resulting from modern cotton aged at greater than 160 °C in air and in nitrogen. Similar strength losses were found but aging in a nitrogen environment did not produce the new carbonyl bands found after aging in air. Thus the underlying aging mechanisms in an air and in a nitrogen environment are not necessarily identical. Losses in strength from thermal and photochemical aging can proceed through nonoxidatiive processes such as loss in degree of polymerization, and changes in polymer crystallinity or hydration. Nonetheless, the correlation between the changes in spectral properties and the loss of tensile strength between artificially and naturally aged cotton textiles suggests that artificial heat aging can be used to approximate the aging processes associated with natural aging (32).

Photolytic Degradation

Photochemical damage in textiles is caused by light-induced chemical changes. The ultraviolet region and shorter wavelength visible regions of the electromagnetic spectra cause most damage. Heat input from high intensity illumination can cause both physical and chemical damage. Temperature and relative humidity are important factors in that the rate of degradation will increase with the moisture content, determined in part by the surrounding relative humidity. Oxygen, present as 21% air, promotes the photooxidation of organic substrates in the formation, reaction, and decomposition of peroxides and hydroperoxides. Studies have shown that the term, "autooxidation" applies whereby oxygen uptake increases with time; and when light is present, the mechanism of degradation is through the initiation, propagation, and termination of hydroperoxides (33). By the mechanism of hydroperoxide formation, various stages of degradation ensue: *inception and induction*, where there may be time to intervene before chain cleavage or crosslinking occurs, the *steady state*, where there is no apparent rate of change in the formation of intermediate compounds and thus a steady state for deterioration has been reached, and a *declining-rate stage* where the limit for reactants consumed and products formed is reached (34). This conceptual framework has been applied to the study of the physical deterioration of organic materials in order to slow, reverse, or impede its development (35). To this effect, various classes of stabilizers have been examined. They include hindered phenols, aromatic amines, sulfur compounds, phosphorous compounds, metal chelates, ultraviolet absorbers, and hindered amine compounds (36).

Biodegradation

Enzymatic degradation, like hydrolytic, thermal, and photolytic, occurs by scission of cellulosic chains according to the kinetics of the reaction. In one case, a fungus cell, *Aspergillus niger*, and a cellulase enzyme from *A. niger*, were applied to heat-aged (to simulate the natural aging process) cotton and flax fabrics (37). In this work, accelerated heat aging was carried out at 105 °C in air, the dark, for time intervals up to 65 days and degradation was evaluated by degrees of polymerization (DP). Comparisons were made to unaged samples. Heterogeneous degradation proceeded from a rapid initial stage, followed by a slower stage. These stages were attributed to the breaking of the anhydroglucosidic (main chain) bonds of cellulose in the amorphous regions. The third stage proceeding at an even slower rate represented the LODP and was attributed to the decreased amount of anhydroglucose bonds. The lower amount of bond breakage in heat aged samples (less than 0.1% compared to 0.4-0.5% for unaged) was attributed to the lower accessibility of heat-aged cellulose because of its carbonyl and carboxylate formation. A comprehensive report on the bioactivity of celllulosic fibers can be found in the work of Vigo (1983) (38). The reader is referred to the reports of Siu (1951) and Becker (1972) for the microbial decomposition of cellulose (39).

Air Pollution

It is recognized that the main gaseous pollutants throughout the world are sulfur dioxide, ozone, nitrogen oxides. Sulfur dioxide is the product of the combustion of coal, petroleum, oil, and natural gas as sulfur residues combine with ambient oxygen in the combustion process. It also originates from natural biological sources. Further oxidation in the presence of water forms sulfuric acid that can promote acidic hydrolysis of cellulosic textiles. Thus ventilation achieved by bringing outside air inside can pose a threat to cotton and flax in the museum environment. Acidity can be neutralized by introducing an alkaline medium into the air stream. Ozone, an oxidizing agent, is recognized as a disinfectant for air and water. However above 2 ppm it is a poisonous pollutant. It forms from the action of ultraviolet radiation on oxygen, from photochemical smog formed from the reaction of sunlight with gaseous exhaust systems, and from certain electrical discharges. As an oxidant of organic material, ozone can react with water to form hydrogen peroxide, a powerful oxidizing material. The nitrogen oxides include N_2O, NO, N_2O_3, NO_2, N_2O_4, NO_3, and N_2O_6. Nitrogen dioxide, an oxidizing agent, is a particular threat to cellulosic substrates because it forms nitric acids in the presence of water and air. In all cases of degradation by pollutants, the relative ratio of amorphous to crystalline regions within a new or aged textile fiber governs the extent of attack and ultimately the rate and extent of degradation. For comprehensive coverage on the pollution effects of gases on dyed and undyed textiles, refer to the work of Upham and Salvin (1975) (40).

Conclusion

The aging properties and characteristics of cellulosic cotton and flax substrates derive from their complex physical and chemical compositions. Where cotton cellulose is unicellular and pure, flax is multicellular and associated with the unique chemical species, lignin, that gives support of stress but which must be discolored to achieved the highly desired white stage. Although the mechanical properties of these two cellulosic substrates are similar with respect to high moisture regain and wet strength, a greater hardness, stiffness, brittleness, and lower work of rupture is characteristic of flax. Similar to other organic materials, cotton and flax fibers are susceptible to loss in chemical and structural integrity when exposed to chemical reagents and microorganisms. This degradation becomes evident by increased brittleness due to selective attack on amorphous regions. The formation of new crosslinks and the development of higher crystallinity impact negatively on mechanical properties and result in loss of hydrogen bonding, oxidation of hydroxyl groups within the main chain and at the hydrolytic terminal or reducing end, acid hydrolysis, photolysis, pyrolysis, environmental exposure, and microbial decay. In the conservation of historic cellulosic artifacts, certain degradation pathways are sometimes evident; for example, the development of yellowness from acid hydrolysis where acidic groups have formed by oxidation over time to form chromophoric groups. Most severe is degradation through aging with the development of oxycellulose compounds that can cause damage to polymer chains, microfibrillar bundles, and ultimate fiber dissolution. This is known for cotton and flax textiles that have been exposed to deteriorating conditions for prolonged periods of time where strength loss is so extreme that handling for conservation treatment may be impeded. Accelerated aging studies of modern cellulosic fabric have shed light on various aging mechanisms that become manifest by the measurement of physical property changes. Whether in an air or in a nitrogen environment, aging progresses with structural attack and chemical modification so that the overall effects obey a kinetic first-order rate profile for the initial stages of degradation before a steady state is reached. Once attained, a steady state for decay is signaled by a leveling off degree of polymerization, thereby indicating severe degradation proceeding at a much slower rate. The application of the empirical rate data for aging, based on a measurable property that represents overall aging effects, can lead to the derivation of the values from which predictions can be made for long-term future effects of treatments designed to slow, impede, or reverse the overall aging process.

References

1. Hollen, N., Saddler, J., Langford, A. L., and Kadolph, S. J. *Textiles;* Macmillan Publishing Company: New York, 1988; pp 24.
2. Tortora, P.G. *Understanding Textiles;* 3rd Edition; Macmillan Publishing Company: New York, 1982; pp 63.
3. Batterberry, M. A. *Fashion the Mirror of History;* Crown Publishers: New York 1977; pp 94.
4. Marsh, J.T., *Self-smoothing Fabrics*, London: Chapman and Hall, Ltd., 1962, pp. 3, 353.
5. Petersen, H., "Cross-linking with Formaldehyde-containing Reactants," in M. Lewin and S.B. Sello (Eds.), *Handbook of Fiber Science and Technology, Vol II. Functional Finishes, Pt.A, Chemical Processing of Fibers and Fabrics*, New York: Marcel Dekker, 1983, Chapter 2.
6. Welch, C.M., "Formaldehyde-free Durable Press Finishes", *Rev. Prog. Color. Rel.* Topics, Vol. 22, 1992, pp. 32-41.
7. Hollen, N., Saddler, J., Langford, A. L., Kadolph, S. J. *Textiles;* Macmillan Publishing Company: New York, 1988; pp 27.
8. Leggett, W.F. *The Story of Linen;* Chemical Publishing Company: New York, 1945; p 17.
9. Leggett, W.F. *The Story of Linen;* Chemical Publishing Company: New York, 1945; p 38.
10. Peters, R.H. *Textile Chemistry*; Elsevier Publishing Company: New York, 1967; Volume 2, pp 250.
11. Peters, R.H. *Textile Chemistry*; Elsevier Publishing Company: New York, 1967; pp 92.
12. Hollen, N.; Saddler, J.; Langford, A. L.; Kadolph, S.J.; *Textiles;* Macmillan Publishing Company: New York, 1988; pp 12-13.
13. Morton, W.E.; Hearle, J. W. S. *Physical Properties of Textile Fibres;* The Textile Institute: London, 1962; pp 288.
14. Morton, W.E.; Hearle, J.W.S. *Physical Properties of Textile Fibres;* The Textile Institute: London, 1975; p. 232, citation (5): Marsden, R.J.B. in *Fibre Science*, (edited by J. M. Preston), The Textile Institute, Manchester, 2nd edition, 1953, p. 229.
15. Textile Research Center, Texas Tech University, Lubbock, Texas. *Textile Topics* Vol. *12*, July, 1984.
16. Hutchins, J.K., *Journal of the American Institute for Conservation* 1983, **22**, 57-61.
17. Nevell, T.P., In *Degradation of Cellulose by Acids, Alkalis, and Mechanical Means*; Nevell, T.P. and Zeronian, S.H., (Eds.); John Wiley and Sons: New York, 1985; pp 223-242.
18. Hersh, S.P.; Hutchins, J.K.; Kerr, N.; Tucker, P.A. *Proceedings of the Conservation and Restoration of Textiles;* The Soluble Components of Degraded Cellulose; Como, 1980; pp 87-98.

19. Kerr, N., Hersh, S.P., Tucker, P. A., Berry, G.M. In *Durability of Macromolecular Materials;* Eby, R.K., Ed.; ACS Symposium Series No. 95; American Chemical Society: Washington, 1979; pp 357-369.
20. Block, I., *Proceedings of the 9th Annual Meeting of AIC*; The Effect of an Alkaline Rinse on the Aging of Cellulosic Textiles; Philadelphia, May, 1981.
21. Feller, R.L. *The International Institute for Conservation of Historic and Artistic Works*; Bulletin of the American Group, No. 2,Vol. 11, pp 44.
22. Lavoie, G.R., *Unlocking the Secrets of the Shroud,* Texas: Thomas Moore Press, 1997.
23. Pellicori, S.F.; Chandos, R. A. *Industrial Research & Development* 1981, p. 1.
24. Kleinert, T.N. *Holzforschung: Mitteilungen zur Chemie, Physik, Biologie, u. Technologie des Holzes* Band 26, Heft 2, 1972, pp. 46-51.
25. Berry, G.M.; Hersh, S.P.; Tucker, P.A.; Walsh, W.K.; *Preservation of Paper and Textiles of Historic and Artistic Value;* Advances in Chemistry Series, No. 164; American Chemical Society, 1977; pp 229-248.
26. Cardamone, J.M.; Brown, P., In *Historic Textile and Paper Materials II: Conservation and Characterization;* Needles, H.L.; Zeronian, S.H., Eds.; Advances in Chemistry Series No. 212; American Chemical Society, 1986; pp 41-75.
27. Arney, J. S.; Jacobs, A. J. *TAPPI,* 1980, 63, pp. 75-77.
28. Cardamone, J.M.; Brown, P., In *Historic Textile and Paper Materials II: Conservation and Characterization*; Needles, H. L.; Zeronian, S.H., Eds.; Advances in Chemistry Series No. 212; American Chemical Society, 1986; pp 41-75.
29. Cardamone, J.M. In *Historic Textile and Paper Materials I: Conservation and Characterization*; Zeronian, S.H.; Needles, H.L., Eds.; ACS Symposium Series, No. 410; American Chemical Society, 1989; pp 239-251.
30. Feller, R.L.; Bogaard, J., In *Historic Textile and Paper Materials II: Conservation and Characterization;* Needles, H.L.; Zeronian, S.H. Eds., American Chemical Society, 1986; pp 330-347.
31. Cardamone, J.M., In *Historic Textile and Paper Materials II: Conservation and Characterization*; Zeronian, S.H.; Needles, H. L. Eds, ACS Symposium Series No. 410; American Chemical Society, 1989; pp 239-251.
32. Cardamone, J.M.; Gould, M.J.; Gordon, S.H. *Textile Research Journal,* Vol. 57, No. 4, 1987, pp. 235-239.
33. Feller, R.L. *Proceedings of the North American International Regional Conference*; "The Deterioration of Organic Substances and the Analysis of Paints and Varnishes," Williamsburg, Virginia, and Philadelphia, Pennsylvania, 1972.
34. Feller, R.L. *Preservation of Paper and Textiles of Historic and Artistic Value*; Williams, J. C., Ed.; Advances in Chemistry Series No. 164; American Chemical Society, 1977.
35. Berry, G.M.; Hersh, S. P.; Tucker, P. A.; Walsh, W.K.; *Preservation of Paper and Textiles of Historic and Artistic Value*; Williams, J. C., Ed.; Advances in Chemistry Series No. 164; American Chemical Society, 1977; pp 248-.260.
36. Rene de la Rie, *Studies in Conservation* 1988, Vol. 33, pp. 9-22.

37. Severs, A. M.; Sora, S.; Nocerino, S.; Testa, G.; Rossi, E.; Seves, A. *Cellulose Chemical Technology* 1998, Vol. 32, pp. 197-209.
38. Vigto, T.L., "Protection of Textiles from Biological Attack," in *Handbook of Fiber Science and Technology: Volume II Chemical Processing of Fibers and Fabrics*, Lewin, M. and Sello, S.B. (Eds.), New York: Marcel Dekker, Inc., 1983, pp. 367-426.
39. Siu, R.G.H., *Microbial decomposition of Cellulose*, New York: Reinhold, 1951.
40. Upham, J.B. and Salvin, V.S., *Effect of Air Polluants on Textile Fibers and Dyes*, EPA-650/3-74-008, North Carolina: Research Triangle Park, 1975.

Chapter 3

Chemical and Physical Changes in Naturally and Accelerated Aged Cellulose

David Erhardt, Charles S. Tumosa, and Marion F. Mecklenburg

Smithsonian Center for Materials Research and Education, Smithsonian Institution, Washington, DC 20560–0534

An understanding of the chemical and physical changes that occur in cellulosic materials is crucial to the preservation of many objects in museums and archives. Decisions regarding care, treatment, and appropriate storage environments are based on their effects on the permanence and condition of the objects. Because many of the changes that occur in cellulose occur too slowly to study easily, accelerated aging conditions such as elevated temperatures are often used to speed up these changes. It is necessary to demonstrate that the changes that occur during such aging are comparable to the changes that occur during "natural" aging in museum collections. Extrapolations based on analyses of the degradation products produced in paper samples aged in a matrix of conditions of temperatures and relative humidities predict that accelerated aging at moderate relative humidities up to at least 90°C should satisfactorily replicate natural aging. Analyses of samples of naturally aged paper and other cellulosic materials bear out these predictions. In addition, important physical properties of naturally aged samples also correspond to those predicted from accelerated aging experiments. The results indicate that it is physically and chemically safe to store cellulosic materials in an environment that varies over a range of moderate relative humidities and temperatures. Within this range, chemical stability is increased at cooler temperatures and lower relative humidities.

The first priority of any museum or archive is the preservation of its collections. The effects of use, wear, and handling can be minimized by fairly obvious measures. The chance of catastrophic damage such as that caused by fire or flood can also be reduced through standard precautions and procedures. Once the risk of immediate damage is minimized, damage caused by slow, long-term degradation processes becomes the primary factor determining the lifetime or permanence of an object. Determining how environmental conditions, preservation treatments, and materials and procedures used in conservation or restoration affect the permanence of objects and their materials is not straightforward. It is difficult to study the slow, long-term degradation processes of natural aging on the laboratory time scale. Observed short-term changes may be too small to measure or extrapolate confidently, and often are not representative of the long-term aging process. One alternative is to study naturally aged materials, but these are usually the result of uncontrolled "experiments" where neither the original state of the starting materials nor the experimental conditions are known. The conditions of "natural" aging are rarely well defined or constant. Another alternative is to try to speed up the natural aging process using exaggerated or extreme conditions such as elevated temperatures, humidities, or light levels. "Accelerated aging" is an attempt to simulate in a short time the effects of long periods of natural aging. Implicit in such experiments is the assumption that the accelerated results are equivalent to those of longer periods of natural aging. Such assumptions are not always true, and even if so are difficult to prove. Our research has established methods for evaluating the relevance of accelerated aging conditions and assessing the effects of aging, both accelerated and "natural".

Experimental

The authors have previously reported methods and results for the analysis of the soluble degradation products of artificially aged paper (1,2). Samples of Whatman #1 paper were aged under a matrix of conditions of temperature and relative humidity (RH) ranging from 50 to 90°C and from 30 to 80% RH for times ranging from 34 to 861 days. Approximately one gram portions of the paper samples were finely divided and stirred in 25 mls of water for two hours. Filtered aliquots of the extract were transferred to weighed reaction vials and evaporated in a vacuum dessicator. The weight of the residue was determined, and 0.1 ml of STOX (3) added per 1.5 ml of residue. STOX is a commercial reagent consisting of 25 mg/ml hydroxylamine hydrochloride in pyridine with 0-phenyl-ß-D-glucopyranoside as an internal quantitative standard. The mixture is heated at 70°C for one hour to convert any carbonyl groups in the residue to the oximes. The same volume of a silylating agent such as hexamethyldisilazane is added, any resulting precipitate is allowed to settle, and the supernatant used for analysis. The derivatized extracts were analyzed by gas chromatography on a 50% diphenyl-, 50% dimethylpolysiloxane capillary column on a Carlo Erba 5300 series gas chromatograph with flame ionization detection and on-column injection at 50 KPa pressure. The initial temperature of 50°C was raised immediately by 10°C/min to 250°C and held for ten minutes. Similar procedures were used in the analyses of naturally aged materials for the present study.

The physical property measurements were conducted using custom-built miniature tensile testers as previously described (4). These testers are capable of making measurements on small samples, and can be manually operated. Equilibrium stress-strain curves are generated under constant environmental conditions by allowing stress relaxation to occur between small incremental increases in strain. Dimensional moisture isotherms are generated by making small changes in the relative humidity and adjusting the gage length to maintain only minimal tension until the specimen equilibrates.

For the present work, naturally aged samples were selected that had no observable damage, such as staining or rot, that would indicate prior exposure to an extreme environment or extraordinary conditions. These included samples of paper from dated books (1804 and 1793), linen thread from the binding of the 1793 book, and samples of wood from a tvihøgdloft (storehouse) in Norway built circa 1650 according to dendrochronological analysis. These samples were analyzed and tested using the same techniques as for the samples in the accelerated aging studies.

Accelerated Aging

It is quite easy to induce changes in materials using exaggerated or extreme conditions, but less easy to prove that the resulting changes are equivalent to natural aging, or more generally, that any two sets of aging conditions are equivalent. The task is further complicated if multiple reactions or processes are taking place, and especially so if the activation energies of the various reactions are similar.

The kinetics of the aging process, and the requirements for different sets of aging conditions to be equivalent, were considered previously by the authors (1,2). Basically, two sets of aging conditions can be considered equivalent if the changes they induce are the same (different amounts of time may be required, of course). This can happen if and only if the same reactions and processes occur under both sets of conditions, and the relative rates of the individual reactions are the same. Accelerated aging, for instance, should speed up all of the reactions of natural aging by the same factor without introducing new reactions. This can be determined most directly by employing methods that can determine the rates of individual reactions rather than just changes in bulk properties. It is difficult to evaluate changes in the aging process using measurements of properties (such as strength or color) that can be affected by more than one reaction. The strength of cellulosic materials, for example, can be reduced either by hydrolysis or oxidation. Once it is established that the chosen accelerated aging conditions are relevant to natural aging, however, any useful property may be used to evaluate the extent or effects of "aging". For example, changes in strength and color are more readily related to the "condition" of paper than is the number of hydrolyzed β-glucosidic bonds, even though such changes may say little if anything about the mechanism or process that produced them.

The authors previously reported the use of analyses of the soluble degradation products of cellulose to compare the effects on paper of aging conditions within a matrix of different temperatures (50-90°C) and relative humidities (30-80%) (1). Figure 1 is a gas chromatogram of the mono- and oligosaccharides and other degradation products extracted from paper aged at 80°C and 77% RH for 278 days. This result is typical of the mixture of degradation products found within the range of environmental conditions of the study, although the distribution of products did vary with the aging conditions and aging time. The predominant products are glucose and its di-, tri-, and tetramers. These are simply short pieces of the cellulose molecule, which is a polymer of glucose, and result from hydrolysis at or near the end of either the original cellulose chain or larger fragments previously split from the original chain. Some of the other, smaller peaks correspond to products produced by oxidation and scission, but it is clear that hydrolysis of the cellulose chain is the primary degradation mechanism during dark aging within the range of temperatures and relative humidities of the study.

This approach allows the study and comparison of the relative rates of individual reactions (as evidenced by the distribution of the resulting products) as a function of the aging conditions. If the relative rates of all of the reactions relative to each other are the same for different conditions, then the aging process is the same and the aging conditions are equivalent. In other words, changing the aging conditions speeds up or slows down all reactions by the same factor. Different conditions may require different lengths of time to reach the same state of aging, but the aging process does not change. If changing the conditions alters the rates of different reactions by different factors, then the aging process is skewed and equivalent states of aging cannot be reached. For example, if a minor product of one aging condition is the major product of another aging condition, then the two aging conditions are not equivalent.

The overall results showed that the aging process of cellulose did not vary dramatically within the range of conditions studied, and that the differences were primarily a function of relative humidity rather than temperature. Figure 2 (from reference 1) shows a detail of the chromatograms for paper aged under varying conditions to the same relative state of aging as judged by comparable amounts of glucose (the main degradation product). The retention time axes are the same for each chromatogram. The vertical scales of the chromatograms are adjusted so that the glucose peak heights are the same (although offscale), so that amounts of each component can be compared directly. The overall aging rates vary with the aging conditions, so different aging times were required to reach the same "state" of aging (similar amounts of glucose). Chromatograms for different temperatures are most similar when the relative humidity is the same. Thus, the aging process could be speeded up without changing it by increasing the temperature but keeping the relative humidity constant. In the same paper, the authors also demonstrated a correlation between chemical and physical changes, with decreases in strength corresponding to increases in hydrolysis products. Hydrolysis is the primary factor in both the chemical and physical aging of cellulose. The most important conclusion of the study was that results from the accelerated aging of cellulose at a controlled relative humidity and temperatures up to 90°C should be relevant to natural aging at similar relative humidities. It must be emphasized that these conclusions apply only to cellulose aged within the range of conditions of the study, and not to other materials or to significantly different conditions. For example, the aging of cellulose exposed to light is different from the dark aging of cellulose as in the work just discussed.

In this paper, we examine the validity of the conclusions drawn from research conducted with accelerated aged paper by comparing predictions based on the previous accelerated aging work with results for naturally aged materials.

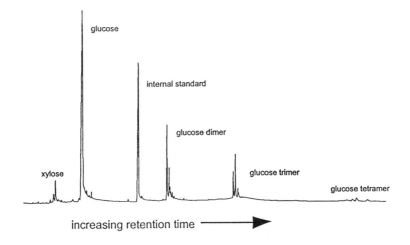

Figure 1. Gas chromatogram of the soluble degradation products in Whatman #1 paper aged at 80°C at 77% relative humidity for 278 days. Saccharides analyzed as the per-trimethylsilylated oximes.

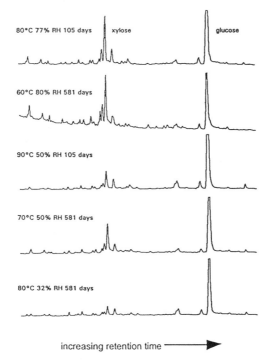

Figure 2. Details of gas chromatograms of the soluble degradation products in samples of Whatman #1 paper aged under a matrix of conditions. Saccharides analyzed as the per-trimethylsilylated oximes.

Natural Aging

With the results of the accelerated aging experiments as a background, samples of naturally aged paper and other cellulosic materials were tested and analyzed using the same techniques (1-2,4-5). The intent was to determine if the chemical and physical characteristics of naturally aged paper correspond to those predicted based on the previous results for accelerated aged samples.

Chemical Changes

Figure 3 is a comparison of chromatograms of extracts of samples of naturally aged and accelerated aged paper. Equivalent retention time ranges are shown for each chromatogram. The vertical scales of the last two sets of ranges are expanded by the same factor for each chromatogram in order to show detail. The naturally aged paper is from the interior of a book printed in 1804, and thus has undergone about 200 years of dark aging. The samples have similar quantities of glucose present, and thus can be considered to be of roughly equivalent "chemical" ages. The types and distribution of reaction products are similar. The chromatogram for the naturally aged paper does have a number of extra peaks corresponding to compounds not present in the accelerated aged paper. Mass spectral analysis of these extra peaks indicated that they are primarily due to degradation products of a gelatin sizing added during manufacture, rather than cellulose degradation products not present in the artificially aged sample. The accelerated aged paper had no sizing. As in the accelerated aged paper, the predominant cellulose degradation products in the naturally aged paper are glucose and its oligomers. The distribution of cellulose degradation products is similar in both samples. Thus, the accelerated aging of the cellulose in a pure cellulosic paper correlates with the degradation process of the naturally aged 1804 paper, even though the 1804 paper contains a non-cellulosic additive. The presence of a small amount of gelatin sizing in the 1804 paper seems to have had little effect on the aging of the cellulose. This is not surprising, since gelatin, like cellulose, is a quite stable material under reasonable conditions. It had been shown previously that cellulose can be quite stable even in a matrix such as wood that contains materials other than cellulose (6).

According to the Arrhenius equation, the ratio of the rates of a reaction at two different temperatures can be expressed as follows:

$$\log \left[\frac{k_2}{k_1} \right] = \frac{E_a}{2.303R} \bullet \left[\frac{1}{T_1} - \frac{1}{T_2} \right]$$

where (k_2/k_1) is the ratio of rates at temperatures T_1 and T_2 (in °K, or °C + 273), E_a is the activation energy of the reaction, and R is the universal gas constant, 1.99cal/°mole. The accelerated aged paper had 147 micrograms of glucose per gram of sample after aging at 80°C (353°K) for 105 days. The 1804 paper had 139 micrograms of glucose per gram of sample after 194 years of natural aging. The ratio of the rates of the two reactions is the inverse of the ratio of the times they take to reach the same state of aging (adjusted for the slightly different concentrations of glucose). Substituting these values along with the previously determined (1) activation energy of 22.7 kcal/mole with $T_2 = 80°C$ (353°K) and T_1 the natural aging temperature yields:

$$\log\left[\frac{\dfrac{147\mu g/g}{105d} \bullet \dfrac{365d}{1yr}}{\dfrac{139\mu g/g}{194year}}\right] = \frac{22,700cal/mole}{2.303 \bullet 1.99cal/°mole} \bullet \left[\frac{1}{T_1} - \frac{1}{353}\right]$$

Solving for T_1,

$$\log(713) = 4960° \bullet \left[\frac{1}{T_1} - \frac{1}{353}\right]$$

$$2.85 = \frac{4960°}{T_1} - 14.05$$

$$16.90 = \frac{4960°}{T_1}$$

$$T_1 = 293° K = 20° C$$

The calculated natural aging "temperature" T_1 is 20.5°C. The book obviously was not aged under constant conditions, having spent at least some time in an attic, a basement, and a used bookstall on the streets of Philadelphia. Nevertheless, deriving an effective aging temperature near the typical temperatures of interior environments provides further support for the correlation between accelerated and natural aging.

Glucose as an Indicator of Age
    Other samples of naturally aged cellulosic materials also have been analyzed. We chose samples with no indication of damage such as staining or rot that would indicate exposure to extreme conditions. In Figure 4, the concentration of glucose for each sample is plotted against its age. (The concentrations of glucose are for the bulk material. No correction was made for the relatively small amounts of non-cellulosic materials such as gelatin or lignin present in some of the samples.) As expected, there is a general increase, due to hydrolysis, in glucose concentration with age. The plot is not linear, probably due to the widely varying nature of both the cellulose sample matrix (paper, linen, wood) and the aging conditions. For example, calculations similar to those above show that the relatively larger concentration of glucose in the 1804 book paper from Philadelphia can be accounted for by assuming that the average temperatures it experienced in Philadelphia were about 4-10°C warmer than the temperatures for the samples from London and Norway.
    One interesting implication of this graph is that the degradation process of cellulose, at least as indicated by the amount of glucose, is relatively independent of

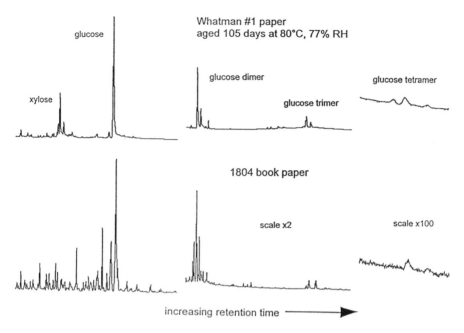

Figure 3. Comparison of gas chromatograms of the soluble degradation products in Whatman #1 paper aged at 80°C, 77% relative humidity for 105 days and paper from a book printed in 1804. Saccharides analyzed as the per-trimethylsilylated oximes.

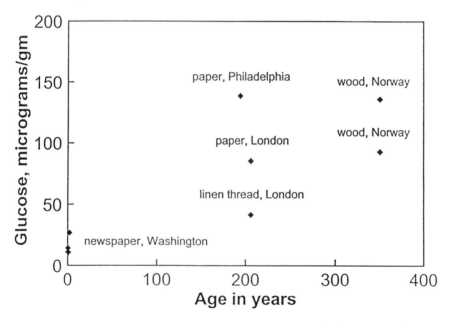

Figure 4. Glucose concentration vs age for naturally aged cellulosic materials.

its source or the matrix it is in (as long as the cellulose is not destabilized by treatment processes or other materials). Further work will refine this model and determine if better correlations between age and glucose content may be obtained by adjusting for known aging conditions, or restricting the plot to samples of similar materials or aging conditions. Such a prospect is illustrated in Figure 5, where glucose concentrations of samples of paper from different dates of The Washington Post are plotted against their age at the time of analysis. From an archivist's point of view, these samples were stored under identical constant conditions, notably the commonly accepted museum "ideal" of 22 C and 50% RH with minimal variation. Evidently, measurable chemical degradation of the cellulose polymer was already occurring even under favorable storage conditions. For these similar samples stored under similar conditions, the plot is linear. This plot, though, predicts much larger amounts of glucose than is found in the older samples. Presumably, the typically acidic nature of newsprint (due to the acid processing of the wood pulp from which it is derived) causes it to degrade at a faster rate than cotton or linen based papers or native cellulose as found in wood (though yielding the same types of products). In this case, the processing does destabilize the cellulose. Thus it will be necessary to take the nature of the sample into account when using the amount of free glucose as an indicator of age. Continuing work will determine the calibration curves that are required. The measurement of glucose concentration provides a sensitive indicator for the process of cellulose decay.

Though yellowed, the newspaper samples were still structurally sound. The yellowing of newsprint is due primarily to the lignin component, but the yellowing of the lignin is a separate process from the hydrolysis of the glucose and is not directly related to changes in strength. The breaking strains of the samples were greater than 2%, indicating that they could tolerate normal handling quite readily.

Physical Changes

The physical properties of naturally aged paper samples also were examined to determine if they were as predicted based on the results from accelerated aging. The evaluation of chemical changes was the primary tool in evaluating the relevance of aging conditions. Once that relevance has been established, appropriate physical tests are required both to validate the interpretation and effects of the chemical mechanisms and to describe the behavior of the cellulosic materials as structural entities. Unfortunately, many of the physical (as well as chemical) tests historically used for paper either do not provide the data required to evaluate the effects of aging or are only minimally useful in assessing the condition of the paper. Many tests are dependent on surface phenomena and not the bulk properties of the paper. Analyzing the types and amounts of degradation products has clarified the effects of environmental conditions on the aging process as well as the relationship between natural and accelerated aging. Measuring the stress-strain behavior of the cellulosic materials has clarified the structural interpretation of the effects of aging. Measurements of the response of cellulosic materials to changes in relative humidity and temperature, and their stiffness, strength, and elastic and plastic behavior are required to calculate the physical response of these materials to handling and preservation treatments as well as environmental influences. This in turn allows the determination of protocols for treatment and handling, and allowable environmental ranges in which no damage or permanent deformations are produced (7,8). Once a physically safe environmental region has been defined, other factors can be considered in deciding whether one area of this region is more suitable or if further restrictions are necessary. Such factors include environmental effects on the rate of chemical degradation processes, the presence of other materials in the collection, phase transitions, the deliquescence of salt contaminants, and the cost, effort, and reliability of controlling the environment within a specified range (8,9).

Stress-strain Behavior of Paper

Figure 6 is a stress-strain curve for the 1804 book paper, in which the stress resulting from an applied strain (dimensional change) is plotted. This curve is similar in shape and magnitude to those for modern and accelerated aged papers, although it ends (the paper breaks) earlier than with unaged paper. An important part of the curve is the initial linear section, which represents reversible (elastic) deformation. If this deformation is not exceeded, there is no damage, either breakage or permanent deformation, to the paper. Aged paper tends to lose its ability to plastically deform (the stress-strain curve ends sooner) but retains its elastic range unless it is so degraded that it has almost no physical integrity. The loss of the plastic region means the paper is more brittle, i.e. it breaks at a shorter extension with less stretching. The ability to plastically deform is less relevant than the retention of the elastic region, however, in determining allowable environmental fluctuations based on reversible behavior. The plastic region becomes more important in considering handling, since a larger plastic region provides a greater range where inappropriate handling or treatment will produce (permanent) deformation rather than breakage. For example, an analysis of the results of classic folding endurance tests indicates that "failure" occurs when the paper is unable to sustain a strain of about 1% (10), which is well beyond the elastic limit. In new paper which has a plastic region extending beyond 1% strain, folding results only in a permanent crease. Old or degraded paper with a breaking strain reduced to below 1% simply fails when folded even if its elastic limit is still intact. Under appropriate environmental conditions and with careful handling, the elastic (reversible) limit should not be exceeded. Information obtained from the stress-strain curves and moisture absorption isotherms (see below) enables one to determine these conditions.

A number of other materials properties can be derived from the stress-strain curve, including the ultimate (breaking) strength, breaking strain, and the modulus (stiffness), which is the slope of the initial section of the curve. Such curves are run under constant environmental conditions. If individual curves are generated for a matrix of environmental conditions, the derived properties can then be determined as a function of the environment. In Figure 7, the moduli (stiffness) of naturally aged, new, and accelerated aged papers are plotted as a function of relative humidity. The older paper is isotropic. The direction of measurement in the anisotropic Whatman paper is indicated. Two important pieces of information derive from this plot; first, the modulus of paper does not change significantly with age; and second, the modulus of paper, new or aged, does not change significantly within the range of

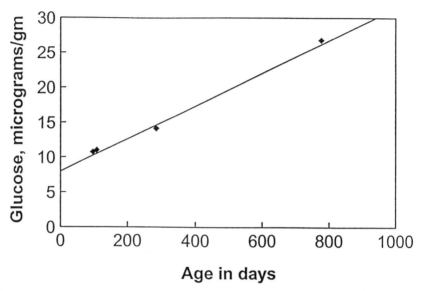

Figure 5. Glucose concentration vs age for newsprint samples from The Washington Post stored at 22°C and 50% relative humidity.

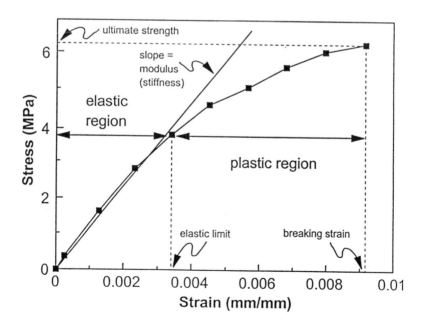

Figure 6. Equilibrium stress-strain curve for paper from a book printed in 1804.

relative humidities that might be considered for display or storage (approximately 20-60% RH). Other significant properties, such as breaking strain and the elastic limit, also were measured and also are relatively unaffected by the value of relative humidity (within this range) at which they are measured. The next question is whether changes of relative humidity within this range can cause damage. If materials are restrained, then forces generated by the differential physical response (expansion and contraction) of materials to environmental changes can be great enough to cause them to exceed the elastic or breaking strains and cause warping or breaking. For example, the cracking of wood veneers can be caused by the differential and directional response to environmental changes of the veneer and its substrate. If a material can respond freely to environmental changes, then no stress is generated. If it is fully or partially restrained from responding by another layer or material to which it is attached that has a different response, then a strain is produced in the material. For this strain to exceed either the elastic or plastic limit, the response to the environmental change of one of the materials must exceed that limit. Environmental changes alone should not cause strains that exceed the elastic limit of paper unless it is completely restrained to a much more environmentally responsive substrate.

In archives and museums, it is crucial to distinguish between the resistance of an object to physical and chemical degradation during "aging" (permanence) and to mechanical damage caused by use (durability). For example, a book may no longer be durable enough to be lent out indiscriminately, but still be quite permanent and retain its value as an object or research tool. In this case, the book has undergone the transition from utile object to icon.

Dimension Isotherms

Figure 8 is a dimension isotherm for the 1804 book paper, in which the change in length of the paper is plotted as a function of relative humidity. Since this paper is isotropic, the change in length (as strain) is plotted for only one direction of the sheet. This isotherm is similar to those for both modern and accelerated aged papers. Note that the relative humidity can undergo extreme fluctuations about the value of 50% RH without causing a dimensional response that exceeds the elastic limit (see Figure 6), which for this paper (and most materials) is about 0.4% (or greater). Compared to many materials used in the construction of books, paper is relatively dimensionally unresponsive to changes in relative humidity. It is in fact the environmental response of materials other than paper that determines the allowable or safe environmental fluctuations in museums and archives. Wood, in the direction tangential to the growth rings, and collagen-derived glues (such as those often found in bookbindings) are both much more responsive to changes in relative humidity. These and other responsive materials restrict allowable fluctuations in museums and archives to about a +/-15% RH range. Even larger variations, especially short term ones, are generally also safe, since these values are calculated using worst-case assumptions such as full mechanical restraint, and immediate and complete changes in moisture content in response to changes in relative humidity. In practice, extremes of relative humidity are rarely maintained long enough for objects to respond fully. It is usually only extreme conditions well outside the recommended ranges (such as central heating without humidification in cold climates) that cause mechanical damage to cellulosic materials. The +/-15% value is appropriate because it is safe and attainable without

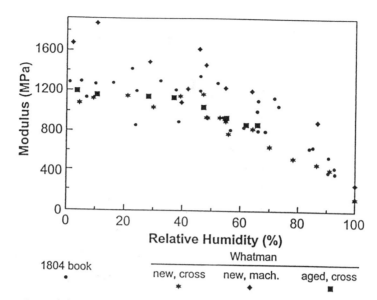

1804 book

Whatman

new, cross | new, mach. | aged, cross

Figure 7. Modulus (stiffness) vs relative humidity for new, accelerated aged (80°C and 77% relative humidity for 105 days), and naturally aged paper. New paper measured in both the machine and cross-machine directions.

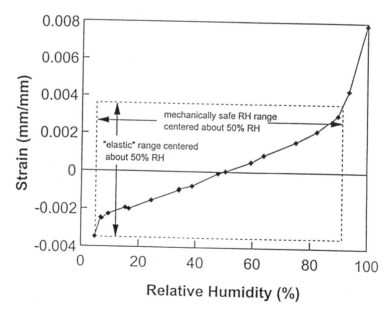

Figure 8. Dimensional moisture absorption isotherm for paper from a book printed in 1804. The allowable range centered about 50% relative humidity in which no irreversible strains can occur even if the paper is fully restrained is indicated.

great expense or specialized equipment. As mentioned earlier, though, other factors must also be considered once a mechanically safe range has been defined.

Reducing the Rate of Deterioration

The purpose of studying the changes that occur during aging is to determine the best approach to eliminating, minimizing, or slowing the rate of these changes.

Cellulose is inherently an exceptionally stable material. Though it can be destabilized (by acid sizes added to some papers, for example), most damage to cellulosic materials results from short term factors such as wear, mold, fire, flood, improper handling, or environmental extremes. Standard precautions and procedures can prevent, minimize, or reduce the chances of such problems. Once physically safe conditions are provided and environmental extremes are avoided, the main concern becomes the reduction of long term deterioration. Within a moderate (30-60% RH) range chosen to avoid mold at high relative humidity and fibrillar collapse and crosslinking at low relative humidity, the rate of hydrolysis varies by a factor of two or three (1), and can be minimized by reducing the relative humidity. Lower temperatures also provide increased chemical stability, and, at least within the human comfort range, introduce no new problems (But see below). Moderate fluctuations of temperature and relative humidity are safely tolerated and have no significant effect on the materials properties of cellulosics (6), so environmental conditions can be modified seasonally to save energy and to avoid the need for specialized or overbuilt environmental controls. Water diffusion is slow enough in bulk or composite structures such as books and wood that even large changes in atmospheric moisture content are effectively moderated. If the environmental and physical storage conditions are properly chosen, the advantages gained from cooler and drier conditions in winter can offset the effect of slightly warmer, moderate relative humidity conditions during summer and provide an increased life expectancy relative to constant "ideal" conditions somewhere in between.

Composite objects such as books contain materials other than cellulose, such as glue, celluloid, leather, and inks. These other components must be considered in developing protocols to minimize the effects of aging. Many of the above considerations apply as well to other materials, especially organic materials, and often for similar reasons. Other objects and materials in a collection, however, may require quite different considerations than those for cellulose. Such considerations have been addressed elsewhere (8,9).

Cool, or even cold, conditions outside the human comfort range are often considered as a way to provide further increases in chemical stability. Lowering the temperature can, in fact, provide increases in chemical stability much greater than those possible by changes in relative humidity (11). For many materials, low temperature conditions are necessary to ensure an adequate life expectancy. However, low temperatures can introduce problems such as brittleness caused by glass transitions and unacceptable changes in moisture content due to temperature dependent moisture absorption isotherms. A complete discussion of such problems is beyond the scope of this article. See reference 10 for a discussion of the considerations related to the cold storage of photographic materials, for example.

Conclusions

Conditions of increased temperature and controlled relative humidity can be used to replicate the natural (dark) aging of cellulose. The types and relative amounts of degradation products in naturally aged cellulosic materials correspond to those seen in accelerated aging experiments, and the rates at which they are produced also correlate well with experiment. Materials properties of naturally aged specimens also are as predicted from experiment. Analyses and tests of both naturally and accelerated aged cellulosic materials can be used to determine appropriate approaches to the preservation of museum and archival collections containing cellulosic materials.

References

1. Erhardt, D.; Mecklenburg, M. F. In *Materials Issues in Art and Archaeology IV;* Vandiver, P. B.; Druzik, J. R.; Madrid, J. L. G.; Freestone, I. C.; Wheeler, G. S., Eds.; Materials Research Society Symposium Proceedings Volume 352; Materials Research Society: Pittsburgh, PA, 1995; pp 247-270.
2. Erhardt, D.; von Endt, D.; Hopwood, W. In *The American Institute for Conservation of Historic and Artistic Works Preprints of Papers Presented at the Fifteenth Annual Meeting, Vancouver, British Columbia, Canada, May 20-24, 1987;* Brown, A.G., Ed.; The American Institute for Conservation of Historic and Artistic Works: Washington, DC, 1987; pp 43-55.
3. Pierce, PO Box 117, Rockford, IL 61105, USA.
4. Mecklenburg, M. F. Ph.D. thesis, University of Maryland, College Park, MD, 1984.
5. Mecklenburg, M. F.; Tumosa, C. S. In *Art in Transit: Studies in the Transport of Paintings;* Mecklenburg, M. F., Ed.; National Gallery of Art: Washington, DC, 1991; pp 173-216.
6. Erhardt, D.; Mecklenburg, M. F.; Tumosa, C. S.; Olstad, T. M. In *ICOM Committee for Conservation 11th Triennial Meeting Edinburgh 1-6 September 1996 Preprints;* Bridgland, J., Ed.; James & James: London, 1996; Vol. 2, pp 903-910.
7. Erhardt, D.; Mecklenburg, M. F.; Tumosa, C. S.; McCormick-Goodhart, M. *WAAC Newsletter.* **1995,** *17(1),* 19-23.
8. Erhardt, D.; Mecklenburg, M. F.; Tumosa, C. S.; McCormick-Goodhart, M. In *The Interface Between Science and Conservation;* Bradley, S, Ed.; British Museum Occasional Paper Number 116; The British Museum: London, 1997: pp 153-163.
9. Erhardt, D.; Mecklenburg, M. F. In *Preventive Conservation: Practice, Theory and Research;* Roy, A.; Smith, P., Eds; The International Institute for Conservation of Historic and Artistic Works: London, 1994; pp 32-38.
10. Tumosa, C. S.; unpublished results.
11. McCormick-Goodhart, M. H. *J. Imaging Science and Technology.* **1995,** *39(2),* 157-162.

Chapter 4

FTIR Study of Dyed and Undyed Cotton Fibers Recovered from a Marine Environment

Runying Chen[1] and Kathryn Jakes[2]

[1]School of Human Environmental Sciences, East Carolina University, Greenville, NC 27858
[2]College of Human Ecology, The Ohio State University, Columbus, OH 43210

Analysis of IR spectra through peak-picking and second derivative calculation methods revealed that while molecular rearrangement was observed in dyed and undyed historic marine cotton and in cotton that had been immersed for three months, the cellulose chemical composition had not been altered by the deep ocean environment. The IR crystallinity index of undyed historic cotton is significantly higher than that of the dyed historic cotton and of two other reference cottons possibly due to biodegradation. These pieces of information allow the conservator to better understand the condition of the material under study, and make better determinations concerning conservation treatment. Increased crystallinity will result in fibers that are less absorbent and less flexible than those with more amorphous character.

Background of the Study

Research concerning marine artifacts has been limited due to the paucity of textiles and other organic artifacts that have been recovered; the choice of techniques employed in their study is limited by the limited sample supply. Knowledge is lacking concerning the chemical and physical structures of marine organic artifacts and of the degradation mechanisms they have undergone (1). In order to preserve the textile artifacts recovered from the deep ocean, fundamental studies of their structures are necessary.

The trunk of textiles recovered from the shipwreck site of the *SS Central America* (*2*), therefore, provides researchers with an unusual opportunity to study the chemical and physical structural changes of the textile materials, and to elucidate the possible degradation processes involved in the marine environment. After the trunk of textiles was recovered from the shipwreck site in 1990, a series of studies has been conducted on particular objects from the trunk, including studies on material treatments, evaluation of the artifacts, fiber structural characterization, and textile fibers' degradation in the marine environment (*3-9*).

This study investigates the chemical and physical structural characterizations of cotton fibers, both dyed and undyed, after a 133 year exposure to the deep ocean environment by employing a Fourier Transform Infrared (FTIR) microspectroscopic technique. An undyed, unbleached plain woven reference cotton fabric (Testfabrics No. 400U) and the same fabric exposed to the deep ocean environment for three months (*6, 9*) were compared with the 133 year exposed cottons for the purpose of exploring the degradation mechanisms that have occurred in the marine environment. FTIR microspectroscopy has the capability of examining micro-samples as small as $10 - 20$ μ in diameter (*10*), and of examining the alteration of chemical structure along the fiber (*11*). The relative crystalline composition of a fiber influences mechanical and sorptive properties of the fiber, such as tensile strength and moisture regain (*12*). For cellulosic fibers, the infrared peak ratio of a $_{2900 \text{ cm-1}}/a_{1370 \text{ cm-1}}$ developed by Nelson and O'Connor's empirical method, was found to be correlated with a variety other crystallinity calculation methods (*13, 14*). The infrared peak ratio was obtained from single fiber infrared spectra as crystallinity indices of the cellulosic fibers under study; the advantage of investigating infrared crystallinity indices from single fiber spectra is that variation in crystallinity between fibers can be studied.

Through the fundamental investigations pursued in this study, determination of the chemical and physical structure of the dyed and undyed cotton fibers provides not only knowledge of long term exposure effects of the deep ocean environment on cotton fibers, but also the essential information required prior to future prescription of any conservation treatments to be applied to historic artifacts from the recovered trunk. The conservation protocol in dealing with textile artifacts requires fundamental knowledge of the chemical and physical condition of the material (*15*).

Methodology

Samples

Four sets of samples were used in this study. Two samples were taken from a man's waistcoat (trunk item inventory number 29178), recovered from the deep ocean *SS Central America* shipwreck site. Preliminary visual examination revealed that the undyed cotton lining of the vest is severely degraded while the brown colored cotton twill of the vest appears to be intact. The two samples, 29178 undyed and 29178 dyed, were labeled MU and MD respectively. For the purpose of comparison, an undyed, unbleached, plain woven cotton, TestFabrics No. 400U (labeled SD), and the same reference cotton fabric that was immersed at the deep ocean site for three months (labeled M3) were included in this study.

Instrument

A Bruker Equinox 55 FTIR microspectrometer with a liquid nitrogen cooled narrow band Mercury Cadmium Telluride (MCT) detector was used in this work. The instrument was purged with nitrogen to reduce the absorption interference of CO_2 and moisture. The operation parameters for collecting cotton fiber infrared spectra were: 4.0 cm^{-1} resolution, 260 or 520 accumulated number of scans, Blackman-Harris 3 - Term apodization function, and zero filling factor 2, with gain switch on. A 15x objective on the microscope was used and the port size of the aperture was 0.3 mm, resulting in a spot size on the fiber of approximately 20 microns in diameter, which is close to the average fiber width in this study, 17 to 23 microns (16). The IR spectra were processed through OPUS IR spectroscopic software (version 2). Further analysis was achieved with Grams 386 (Galactic Industries) and Spectrum 2000 (Perkin Elmer) software packages.

Sample Preparation

During the sample preparation it could be seen that the undyed cotton lining of the vest (MU) was so weak that the yarns broke into small fragments when they were dissected from the small fabric fragment that had been removed from the garment. The fibers removed from the dyed yarns (MD) maintained their integrity as they were separated from the fabric fragment. As a result of the difference in the character of the fibers to be examined, individual fiber samples used for FTIR experimentation were mounted on two types of sample holders. All of the sample fibers were first flattened with a micro-roller device on a glass slide. The slide and the roller were cleaned between samples by spraying with canned air to avoid sample contamination. Because of their fragility, the undyed fibers were each placed on a 13mm KBr window located on a rectangular infrared compression cell holder. The holder allowed manipulation of sample orientation. Because of their greater structural integrity, fibers removed from the other three fabric samples (MD, SD, M3) were mounted to cross a 2 mm slit in a heavy cardboard frame. The fibers were attached to the frame on either side of the slit using double-sided tape. Before beginning the spectrum collection, the samples were kept in a desiccator for over 24 hours in order to reduce interference from the absorbed water in the fiber. All infrared spectra were collected with the fibers oriented in the same vertical orientation with respect to the microscope optics (17).

Two problems are often encountered in the examination of single fibers. One is the diffraction effect that arises from samples of very small size, especially samples under 10 microns. It has been argued, however, that although spectrum quality is affected by diffraction, sample differences are still detectable (*17*). The second problem encountered is the effect of sample thickness and refraction by the sample when transmission or absorbance spectra are collected. It has been suggested that a thickness of 10-20 μ should be obtained (*17*). The usual solution for reducing sample thickness is to press the sample, such as by using a compression cell or a diamond cell. Studies on polymer fibers, however, have indicated that pressing may affect the crystalline structure of the fiber, changing peak intensity and peak intensity ratio (*18*). Because the fibers studied in this work had an average fiber width from 17 to 23 microns, the diffraction effect and absorption problem was not a major concern. Due to the natural non-uniformity between cotton fibers, 45 to 71 infrared spectra were collected from multiple locations of multiple fibers removed from each of the four cotton fabric samples for the purpose of statistical analysis.

Data Analysis

The baselines of the original IR spectra of the cotton fibers showed varying slopes due to the different fiber thicknesses and widths. It was therefore necessary to apply baseline corrections to the spectra before the identification of peaks. A multiple-point baseline correction method in the computer software was applied to absorbance spectra. Transmittance spectra are presented in this report, because this is the form typically employed in the literature on cellulosic fibers. After baseline correction, peaks were identified by employing the "Peak Picking" routine offered in the software package and using a threshold of 2 % of the highest peak intensity. The infrared spectra of the four samples (MD, MU, M3, SD) were compared in each of the absorption regions based on the original spectra and the peak-picking table.

In addition, second derivatives of each of the infrared spectra were calculated for further spectral comparisons. The advantage of second derivative spectrum calculation is that peaks with shoulders and peaks that overlap can be attenuated for easier identification. The mathematical foundation of the second derivative function results in the sharpening of the narrow bands in the spectrum and the flattening of the broad background (*19, 20*). Broad peaks also will be flattened instead of being sharpened. Earlier studies on natural cellulose fibers by using a derivative infrared spectroscopy method showed that shoulders and overlapping peaks can be easily resolved by recording the derivative spectra through an attached differentiator (*21, 22*). In the derivative method reported in the literature, scan speeds and time constants were controlled to obtain the optimum second derivative spectra. In comparison, the method employed in this study is a computer program that calculates the derivative spectra from the digitized data stored in the computer or on the disk. The commonly used method of smoothing and differentiation is based on the least squares procedures which was first developed by Savitzky and Golay (*23*) and later improved by other researchers (*24*). The Savitzky-Golay method, therefore, was chosen to calculate second derivative spectra in this study by choosing the 5 points smoothing function. The peaks thus identified from the calculated derivative spectra were compared with those reported in the reviewed literature (*21,25 - 27*). In this study second derivative spectra were calculated from transmittance data since only peak identification was of interest. Second derivative calculation of absorbance data is required when quantitative analysis is of concern.

In calculating the infrared crystallinity ratio, the relative peak intensity at 2900 cm^{-1} and 1372 cm^{-1} was measured from the infrared spectra by following the method developed by Nelson and O'Connor (*13*). These authors found that the IR bands at 1372 cm^{-1}, 1335 cm^{-1}, and 1315 cm^{-1} were the most sensitive to cellulose decrystallinization realized by ball-milling at various time intervals. The 1372 band, assigned as CH deformation, was chosen as the measure of cellulose crystallinity because the intensity of this band was neither affected by the absorbed water nor by the cellulose lattice type (cellulose I and cellulose II). In order to account for the variations between samples, the 2900 cm^{-1} band was chosen as an internal reference and the relative absorption ratio $a_{1372\ cm-1}/a_{2900\ cm-1}$ was calculated as the infrared crystallinity index (*13, 14*). In measuring the relative band intensity, local baselines of the two bands were drawn on the spectra printouts, then the vertical distance from the peak of the band to the local baseline was obtained. In recent years, others researchers have explored other measures of cellulose crystallinity based on infrared data (*28, 29*). However, because these methods are limited in either justification of band choice or in applicability to fibers, they were not used in this research and the method of Nelson and O'Connor was employed in this study. Different from other studies, the infrared crystallinity indices calculated in this study were obtained from a number of single fiber infrared spectra in order to perform statistical analyses of the results. For each sample, the mean and standard deviation of the IR crystallinity indices were calculated. Then, the studentized T-test and Tukey's pairwise comparisons were employed to find if there were significant differences between each pair of the four samples.

For multiple comparison analysis, the Tukey's pairwise comparison method is capable of indicating any significant different between any pairs of the samples or groups. One-way analysis of variance (ANOVA) can indicate if a significant difference exists among the samples but no information is provided on where the significant difference exists. In comparison with other pairwise comparison methods, Tukey's method is acknowledged as a more vigorous method in error control (*30, 31*). In this study, Tukey's pairwise comparisons were conducted employing Minitab version 10 statistical software. However, Tukey's procedure is challenged when different sample size and variance are included in the family of comparison. To address this concern, the two tailed studentized T-test with alpha = 0.05 was conducted, though it is usually used when comparing only two groups or samples because with each additional comparison the alpha level is reduced. The level of conservatism and liberalism in error control can therefore be compared by providing two sets of statistical results.

Results

Infrared Spectra Comparison

The infrared spectra obtained along each single reference cotton fiber and those along each single immersed reference cotton fiber showed little variation in their baselines, peak positions, and peak intensities. However, variations were observed among the spectra scanned from different spots on each single 29178 dyed and undyed cotton fiber. These differences coincided with the morphological variations of the spots being scanned, such as the fiber wall thickness and the dark deposits observed at some of the spots. When very dark spots were examined, the IR spectrum was observed to have a positive slope baseline due to the thickness and high absorbency of the deposits. The baselines of these spectra, therefore, were corrected from infrared absorbance data prior to further spectral analyses. The spectra from areas that were free of deposits but were otherwise degraded did not display observable differences from those locations that were apparently undegraded. The implication of this similarity is that no observable chemical alteration can be found from the normal infrared spectra of these different cotton fiber spots, whether visibly degraded or undegraded.

The IR spectra collected from different fibers of the same sample showed differences related to physical features, such as fiber wall thickness and flat or convoluted fiber surface. These differences could be observed at the OH stretching peak around 3300 cm^{-1}, the CH stretching peak around 2900 cm^{-1}, and the coupled band area between 1200 cm^{-1} to 1000 cm^{-1} (see Figure 1). Distinctly different IR spectra were obtained from very thin walled immature cotton fibers. These spectra are characterized by a very sharp OH stretching peak and sharp peaks of higher relative intensity in the 1200 - 1000 cm^{-1} region in comparison with the spectra collected from normal mature cotton fibers (see spectra b and d in Figure 1). The increased sharpness along the spectra can be attributed to the thinness of the fiber which thereby reduces band broadening. The fiber thickness is equivalent to the optical pathlength in Beer's Law. This law states that absorption increases with increase in the optical pathlength. The decreased sharpness and intensity of the CH stretching peak in the immature cotton fiber relative to a mature cotton fiber is related differences in their crystalline structures. An earlier study of crystallinity determined by x-ray diffraction found that phase transformation from cellulose I to cellulose II upon NaOH treatment was directly related to maturity of the cotton sample (32). Another study found that shoulders and peak sharpness in the CH stretching region of cotton fibers were reduced when the fibers were deformed by pressing at high pressure (33). The pressing not only reduced the fiber thickness but also altered the fiber's crystalline structure as shown in the differences in the calculated second derivative spectra between the pressed and unpressed fibers. The same study also showed that the spectra of the pressed cotton fiber was characterized by sharpened peaks in the OH stretching region and the region between 1200 cm^{-1} and 1000 cm^{-1}.

44

Peak Identification

In this research infrared peaks were first identified from the normal infrared spectra by setting the transmission threshold at 2 per cent. Then second derivative spectra were calculated using a 5 point smoothing function. Most of the peaks thus identified were consistent among the four samples within a range of ±2 cm⁻¹. These peaks are also close to the peaks identified in earlier studies by other researchers, within approximately ±2 cm⁻¹ range.

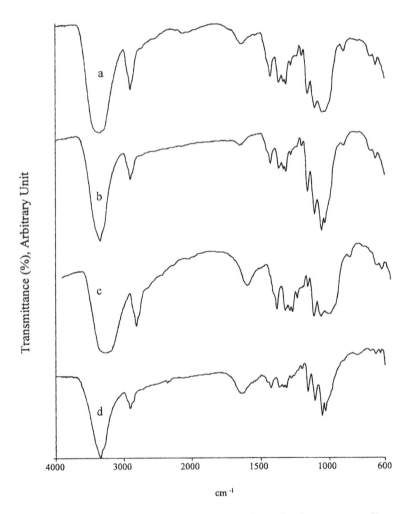

a = normal reference cotton fiber.   b = thin-wall reference cotton fiber.

c = normal fiber 29178 dyed cotton.   d = thin-wall fiber from 29178 dyed cotton.

Figure 1. Infrared transmittance spectra of thin-wall cotton fiber and normal cotton fiber.

The OH Stretching Region

A large broad OH stretching peak is observed from most of the cotton fibers with normal secondary wall thickness. The band broadening in this region can be attributed to both the sample thickness and the presence of absorbed water. Immature cotton fibers display a sharp peak in this region. Due to the broad peak in the OH stretching region, second derivative analysis did not reveal different OH vibration modes for most of the fiber locations examined. Earlier studies proposed that the OH stretching region contains different vibration modes of the six (parallel-chain model) or 12 (anti-parallel model) OH groups in each unit cell of cellulose I, and these OH groups are involved in either intermolecular or intramolecular hydrogen bonding, or both (*25, 26*). A number of vibrational modes in this region were reported in an earlier study using a derivative infrared spectroscopy method (*21*).

The CH Stretching Region

The CH stretching region in most of the cotton fibers appears as a sharp peak at 2901 cm^{-1} with several shoulders. Through second derivative analysis, all of the same CH vibration modes identified by polarized infrared methods in earlier studies were observed (Figure 2). The peaks at 2969 cm^{-1} (CH stretching) and 2945 cm^{-1} (CH2 anti-symmetric stretching) were the strongest and were consistently identified in the four cotton samples. Other peaks, 2915 cm^{-1}, 2911 cm^{-1}, 2901cm^{-1}, 2869 cm^{-1}, and 2850 cm^{-1} were also identified. In comparison, the bands identified in Pandey's (*21*) study in this region are 2967cm^{-1}, 2941cm^{-1}, 2915 cm^{-1}, 2890 cm^{-1}, 2874 cm^{-1}, and 2849 cm^{-1}. Similarities and differences are found between the two. This may be due to the different sample preparation method rather than to the method of obtaining second derivative spectra.

The Region from 1500 cm^{-1} to 1200 cm^{-1}

The peaks in this region are due to various CH and OH bending modes. Though most of the peaks were identified by the peak-picking procedure, additional peaks were identified by the second derivative method (Figure 3). Peaks identified in this region by Laing and Blackwell were identified in this study as well. In addition a peak at 1462 cm^{-1} was detected in all samples but the reference cotton. No 1482 cm^{-1} peak was detected in this sample either. Two peaks, at 1429 cm^{-1} and 1422 cm^{-1}, were identified in spectra of all the four cotton samples. In the region between 1400 cm^{-1} and 1200 cm^{-1} the peaks identified in all four samples were consistent with those described in the literature. The only exception is the modern cotton fibers for which a peak at 1360 cm^{-1} next to the peak at 1372 cm^{-1} was identified, corresponding to the weak peak at 1360 cm^{-1} identified by Blackwell. In comparison, Pandey found fewer peaks in this region, but peaks between 1700 cm^{-1} and 1500 cm^{-1} were identified. The different peaks identified by Pandey include 1449 cm^{-1}, 1299 cm^{-1}, and 1227 cm^{-1}.

The above peak identification results indicate consistent differences between the modern cotton fiber in comparison to the other three cotton samples, suggesting that the cellulose structure of the fibers exposed to the deep ocean environment was modified by swelling. Infrared peak intensity is related to the concentration of the absorbing chemical moieties. As the compositions of the isomers that give rise to particular absorption peaks at particular wavenumbers are altered by molecular rearrangement, the infrared spectra change (34). In the marine environment the cellulose fibers are swollen, water molecules and other ions from the seawater enter the accessible regions, inter and intra molecular hydrogen bonds are interrupted and

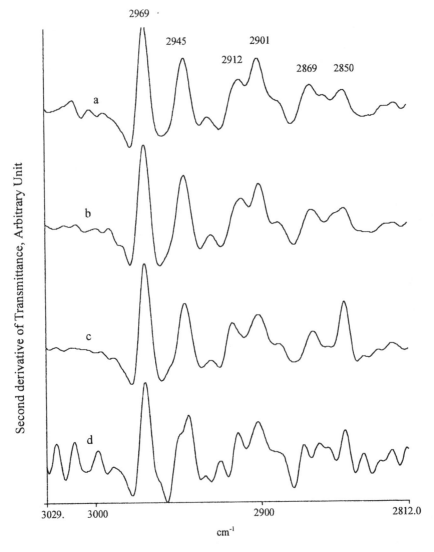

a = reference   b = three-month immersed
c = 29178 dyed   d = 29178 undyed

Figure 2. Second derivative infrared spectra of CH stretch region.

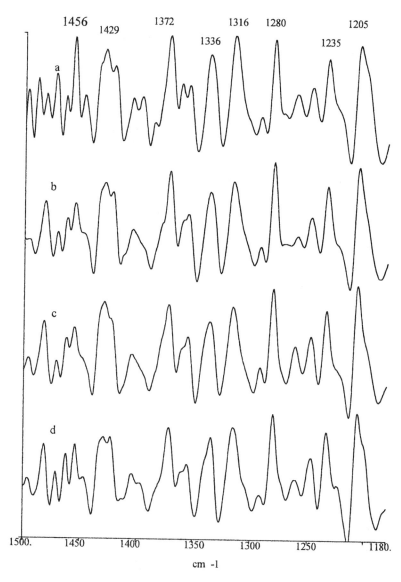

a = reference cotton   b = three-month immersed cotton

c = 29178 dyed cotton   d = 29178 undyed cotton

Figure 3. Second derivative infrared spectra in the region from 1500 to 1200 cm-1.

the cellulose molecules can rearrange to some extent, thus releasing internal strains. As new hydrogen bonds are established between molecules in these disordered regions, certain molecular isomers increase and the corresponding infrared absorbance intensity also increases. This structural modification may give rise to the appearance of the two peaks at 1482 cm^{-1} and 1462 cm^{-1} in the swollen cotton fibers, and the disappearance of the peak at 1360 cm^{-1}. This proposition is partially supported by the appearance of a peak at 1364 cm^{-1} in cotton fibers treated with NaOH.

The Region from 1200 cm^{-1} to 900 cm^{-1}

Peak assignments are difficult in the region 1200 - 900 cm^{-1} due to the coupling between different vibration modes. Earlier studies assigned the peaks in this region to different vibration modes involving C-O stretching. All the four cotton samples showed the greatest variation in both peak shapes and their relative intensities in this region. The thin walled cotton fibers, however, showed sharp peaks of higher relative intensity (Figure 1). In addition to the peak assignments in this region reported in earlier studies, two more peaks were consistently identified from three cotton samples with the exception of modern cotton fiber, one at 1103 cm^{-1} and the other at 1046 cm^{-1} (Figure 4). The two peaks are close to the peaks reported by Blackwell at 1098 cm^{-1} and 1043 cm^{-1} (26). This result again shows the difference between the modern cotton sample and the other three cotton samples.

The Region From 950 cm^{-1} to 600 cm^{-1}

Only a few peaks were identified in this region. A peak located in the range from 896 cm^{-1} to 900 cm^{-1} was identified in all four samples. A shoulder at 800 cm^{-1} assigned as one of the ring breathing vibration modes was identified only in the spectra of the two marine cotton samples. In the region below 800 cm^{-1} great variation was observed not only between samples but also between different fiber specimens from the sample. No peaks in this lower wavenumber region were reported in Pandey's study.

Infrared Peak Ratio Crystallinity Indices

Nelson and O'Connor's method has been referred widely in the literature due to the fact that a large number of samples were compared through a number of crystallinity measurement methods (14, 35). The correlation coefficients between the infrared crystallinity ratio index proposed by Nelson and O'Connor and all other methods fell in the range from 0.92 to 0.86 (13, 35). In the research reported herein, the relative peak intensities at 2900 cm^{-1} and 1372 cm^{-1} were determined from each of 45 to 70 IR spectra of each sample, and the ratios of intensities were calculated. The large number of spectra was collected for the purpose of insuring statistical representation of the samples. The average values of the measured crystallinity indices from the four cotton samples are smaller than those reported in the literature. These literature studies differed from present work in that a KBr pellet method for

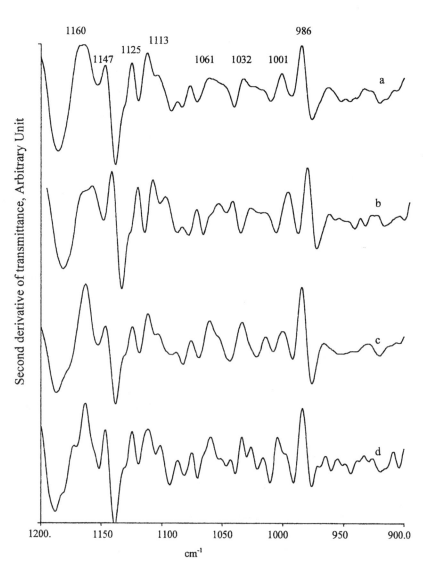

a = reference cotton   b = three-month immersed cotton

c = 29178 dyed cotton   d = 29178 undyed cotton

Figure 4. Second derivative infrared spectra in the region 1200 - 900 cm-1.

sample preparation was employed along with a much larger sample size. In those reports the fibers in the pellet were first chopped and then pressed along with the KBr powder under relative high pressure. For example, the infrared crystallinity index of purified cotton obtained in earlier studies was 0.9 (*13, 36*), while the values determined in this work ranged around 0.6. The difference in crystallinity is likely due to the differences in sample preparation. Comparison between samples prepared in an equivalent manner, however, provides an indicator of relative crystallinity which can be used to compare one fiber to another.

In analyzing the resultant data, descriptive statistics were obtained first as listed in Table I-A. The standard deviation of the undyed cotton fibers' crystallinity indices is larger than that of the other samples, therefore some additional spectra were collected from this sample. Three statistical tests were employed.. First, one-way analysis of variance (ANOVA) was conducted. It was found that significant differences exist between the four samples. Studentized T-test comparisons were conducted between each possible pair of the four samples. The T-test can be a powerful statistical procedure for paired comparisons but this test does not control for the familywise error rate and can result in an increased possibility of a Type I error. The T-test results listed in Table I-B show that four pairs of means that are significantly different: SD-MU, M3-MD, M3-MU, and MD-MU.

Tukey's pairwise comparisons were conducted using the procedure for unequal sample size and heterogeneous variances between samples. In comparison with other pairwise comparison methods, such as Hsu's Multiple comparison with the best test or Dunnett's procedure, the critical value is small, reflecting a test with lower power but less potential for type I error. The pairwise comparison results among the four samples' crystallinity indices are listed in Table I-C. The outputs are pairs of intervals between the two members which was obtained by calculating the family mean and the member mean while the family error rate was controlled. When the two intervals have the same negative or positive sign, the two means are significantly different. Only two pairs of means were found significantly different: SD-MU and MD-MU. Therefore, a conservative judgment with less possible type I error reveals that the mean crystallinity index of the undyed cotton fiber differs significantly from the reference cotton and from the dyed cotton from the same environment. On the other hand, short term immersion has no significant effect on the infrared crystallinity index of reference cotton.

According to the earlier studies (*14, 29*) the value of the infrared crystallinity index increases with increased crystallinity of cellulose I. Therefore, the crystallinity index of 29178 undyed cotton fiber, 0.66 ± 0.21, compared to the index of the reference cotton, 0.57 ± 0.09, indicates an increased relative crystallinity. Research on the enzymatic degradation of cellulose, one of the possible causes of cellulose degradation, showed a slight increase in crystallinity which may be due to the removal of the molecules in the amorphous region after enzyme attack (*14, 37*). Thus the severe biodegradation observed in the microscopic examination of 29178 undyed fibers is the possible cause of the increased relative crystallinity of the remaining fibrous material. Bacteria degrade cellulose by producing enzymes that cause cellulose depolymerization; the absorption and reaction of these enzymes would

Table I. Summaries of Statistical Analyses of Infrared Crystallinity Indices

Table I-A	Descriptive statistics			
	Sample	N	Mean	S-dev
	SD	50	0.57	0.09
	M3	50	0.60	0.10
	MD	45	0.54	0.08
	MU	71	0.66	0.21

Table I-B	Returned T-test probability of per pair comparison (Two-tailed test, $\alpha = 0.05$)		
	SD	M3	MD
M3	0.151		
MD	0.089	**0.002**	
MU	**0.001**	**0.021**	**0.000**

Table I-C	Tukey's pairwise comparison: Intervals between family mean and member mean FER = 0.05, IER = 0.01, Critical value = 3.36		
	SD	M3	MD
M3	-0.101 0.045		
MD	-0.045 0.105	-0.017 0.133	
MU	**-0.162** **-0.027**	-0.134 0.001	**-0.194** **-0.055**

Legend: SD=reference cotton, M3=three-month immersed reference cotton, MD=29178 dyed cotton, MU=29178 undyed cotton. FER = family error rate, IER = individual error rate.

occur most readily in accessible amorphous regions of the fibers being attacked (*14*). Another source of this large variance in crystallinity indices of 29178 undyed cotton fibers may be due to interference in the spectra arising from the contamination observed on the fiber. However, because very similar contamination was observed on both the 29178 dyed and undyed cotton fiber and the standard deviation of the infrared crystallinity index of the dyed fibers is much smaller than that of the undyed cotton fibers, localized biodegradation is a more likely factor contributing to the crystallinity changes among the undyed cotton fibers from the marine environment. The significant difference in crystallinity indices between the dyed and undyed cotton fibers indicates that while it is possible that the two fibers are of different cotton origins, the effect of long term exposure to the deep ocean environment varies as the fibers' physical or chemical states differ, and the presence of dye and mordant contained in the dyed cotton fiber resulted in a different pathway of degradation (*16*).

Conclusion

The infrared spectra collected from the different spots along the same fiber are very similar to each other, indicating both physical and chemical structural continuity along the fiber. The spectra collected from the immature fibers show sharp OH stretching peaks and more defined and intense peaks in the 1200 - 1000 cm^{-1} region. Calculation of the second derivative was effective in revealing most of the cellulose I IR peaks comparable to those identified in earlier studies reported in the literature. However, some differences were found, which may be attributed to the difference in sample preparation but also to the differences in the cellulose samples used in this research. No fundamental chemical alteration in the dyed or undyed cotton fibers exposed to the deep ocean environment was detected by the second derivative single fiber FTIR method. However, cellulose molecular rearrangement was revealed by the spectral variation observed between the modern cotton and the other three cotton samples with marine environment history. In contrast, cellulosics that have experienced long term storage in other environments sometimes display products of oxidation or hydrolysis, such as carbonyl and carboxyl groups (*38*). The fact that the marine fibers exhibit no increase in oxidative or hydrolytic products allows the conservator to better understand water-logged or wet site artifacts which have experienced a different degradation path than materials of the same age stored in a dry environment.

Not only is the mean value of the crystallinity indices of the undyed historic cotton significantly higher than those of the other three cotton samples, the indices also display a larger variation in values than do the other groups. The most likely cause is the severe biodegradation incurred by the undyed cotton in the deep ocean environment. The dyed cotton fiber, removed from the same garment as the undyed cotton, did not exhibit biodegradation and the variation in crystallinity indices was similar to that of the other two reference cottons. The differences between the dyed and undyed cotton fibers can be attributed to the chemical and physical differences between them, such as the dye and mordant contained in the dyed cotton fiber.

Increased crystallinity will result in fibers that are less absorbent and less flexible than those with more amorphous character. This information provides the conservator with knowledge that the conservation protocol should consider the differences between fabrics from the same object, and that different treatments may be needed for the two types of cotton fabric from the garment.

The cotton fiber immersed for a short period did not reveal significant differences from the untreated reference cotton. The mean values and variation in crystallinity indices of the immersed and unimmersed reference cotton are similar to each other, indicating that the short term exposure of three months did not alter the physical structure of the cotton fiber. Studies on additional materials exposed for longer periods than three months is needed to provide more information towards the pathways by which the fibers are altered in the deep ocean.

The research reported herein also has application to the study of cellulosic materials other than those exposed to the marine environment. The techniques of peak-picking and second derivative calculation provide data concerning the infrared absorbance of single fibers that can be used in understanding the chemical and physical condition of the materials.

References

1. Florian, M-L. E. In *Conservation of Marine Archaeological Objects*, Pearson, D. Ed.; Butterworths: New York, 1987; pp 21-54.
2. Herdendorf, C. E.; Thompson, T. G.; Evans. R. D. *Ohio J. Sci.* **1995**, *95*, 4.
3. Jakes, K. A.; Mitchell, J. C. *J. Am. Institute for Conservation.* **1992**, *31*, 343.
4. Crawford, L. C. Ph.D. thesis, The Ohio State University, Columbus, OH, 1994.
5. Hannel, S. L. *The Acquisition of Menswear in An American Frontier Town, 1857*; M. S. Thesis: The Ohio State University, 1994.
6. Jakes, K. A.; Wang, W. *Ars Textrina* **1993**, *19*, 161.
7. Foreman, D. W.; Jakes, K. A. *Textile Res. J.* **1993**, *63*, 455.
8. Srinivasan, R.; Jakes, K. A. In *Materials Issues in Art and Archaeology V*; Vandiver, P.; Druzik, J.; Merkel, J. F.; Stewart, J. Eds.; Materials Research Society Symposium Proceedings, Vol. 462; Pittsburgh, 1997; pp. 375-380.
9. Wang, W. Master's thesis, The Ohio State University, Columbus, OH, 1992.
10. *Infrared Microspectroscopy*; Messerschmidt, R. G.; Harthcock, M. A., Eds.; Marcel Dekker, Inc.: New York, 1988.
11. Wetzel, D. L. *USA Microscopy and Analysis* **1996**, *No.18*, 17.
12. Morton, W.E.; Hearle, J.W.S. *Physical Properties of Textile Fibers*; Textile Institute: Manchester, 1993.
13. Nelson, M. L.; O'Connor, R. T. *J. Appl. Polymer Sci.* **1964**, *8*, 1325.
14. Krässig, H. A. *Cellulose: Structure, Accessbility and Reactivity*; Gordon and Beach Science Publishers: Philadelphia, 1993.
15. Crighton, J.S. In *Polymers in Conservation*, Allen, N.S.; Edge, M.; Horie, C.V. Eds.; Cambridge: The Royal Society of Chemistry, 1992; pp.83-107.
16. Chen, R. Ph.D thesis, The Ohio State University, Columbus, OH 1998.
17. Katon, J. E.; Sommer, A. J. *Anal. Chem.* **1992**, *64*, 931A.

18. Tungol, M W.; Bartick, E. G.; Montaser, A. *Appl. Spectrosc.* **1990**, *44*, 543.
19. Whitbeck, M.R. *Appl. Spectrosc.* **1981**, *35*, 93-95.
20. Maddams, W.F.; Southon, M.L. *Spectrachim, Acta* **1982**, *38*A, 459-466.
21. Pandey, S.N. *J. Appl. Polym. Sci.* **1987**, *34*, 1199-1207.
22. Pandey, S.N. *Textile Res. J.* **1989**, *59*, 226-231.
23. Savitzky, A.; Golay, M.J.E. *Anal. Chem.* **1964**, *36*(8), 1627-1639.
24. Madden, H.H. *Anal. Chem.* **1978**, 50(9), 1383-1386.
25. Liang, C. Y.; Marchessault, R. H. In *Instrumental Analysis of Cotton Cellulose and Modified Cotton Cellulose*, O'Connor, R. T. Ed.; Marcel Dekker, Inc.: New York, 1972; pp.59-91.
26. Blackwell, J. In *Cellulose Chemistry and Technology*, J. C. Arthur, Jr. Ed.; ACS Symposium Series 48, American Chemical Society: Washington, DC., 1977; pp.206-218.
27. Fengel, D. In *Cellulosics: Chemical, Biochemical and Material Aspects*, Kennedy, J. F.; Phillips, G. O.; Willians, P. A., Eds.; Ellis Horwood: New York, 1993; pp.135-140.
28. Hulleman, S.H.D.; van Hazendonk, J.M.; van Dam, J.E.G. *Carbohydr. Res.* **1994**, *261*, 163-172.
29. Schultz, T.P.; McGinnis, G.D.; Bertran, M.S.J. *J. Wood Sci. and Tech.* **1985**, *5*(4), 543-557.
30. Toothaker, L.E. *Multiple Comparison Procedures*; Sage Publications, Inc.: California, 1993.
31. Hochberg, Y.; Tamhane, A.C. *Multiple Comparison Procedures*; John Wiley & Sons: New York, 1987.
32. Kulshreshata, A.K.; Chudasama, V.P.; Dweltz, N.E. *J. Appl. Polym. Sci.* **1975**, *19*, 115-123.
33. Chen, R.; Jakes, K. Presentation at the Central Region American Chemical Society meeting, 1999, June 23. Columbus, OH.
34. Koenig, J.L. Spectroscopy of Polymers; American Chemical Society: Washington, DC., 1992.
35. Tripp, V.W. In *Cellulose and Cellulose Derivatives* Part IV, Bikales, N.M.; Segal, L. Eds. Wiley-Interscience: New York, 1971; pp.305-323.
36. Pandey, S.N.; Zyengar, R.L.N. *Textile Res. J.* **1968**, *38*, 675.
37. Schurz, J.; Billiani, J.; Hönel, A.; Eigner, W. D.; Janois, A.; Hayn, M.; Esterbauer, H. *Acta Polymerica* **1985**, 36, 79.
38. Cardamone, J.M.; Keister, K.M.; Osareh, A.H. *Polymers in Conservation*; Royal Society of Chemistry:Cambridge, 1992.

Acknowledgments
    The authors wish to thank Dr. Terry Gustafson, Department of Chemistry, the Ohio State University for his insightful comments during the performance of this research. The support of the Columbus-America Discovery Group and that of the Ohio State University Graduate School are gratefully acknowledged. The authors also appreciate the suggestions and recommendations made by anonymous reviewers of the draft document.

Chapter 5

Characterization of Chemical and Physical Microstructure of Historic Fibers through Microchemical Reaction

Runying Chen[1] and Kathryn A. Jakes[2]

[1]**School of Human Environmental Sciences, East Carolina University, Greenville, NC 27858**
[2]**College of Human Ecology, The Ohio State University, Columbus, OH 43210**

Light microscopy and FTIR microspectroscopy were employed to study the physical and chemical structural changes of dyed and undyed historic marine cotton, and comparative reference cottons when treated with alkaline solutions. After treatment with 18% NaOH, the extent of swelling differed but the chemical structures of all four specimens were similar. After treatment with a mixed solution consisting equal parts of 18% NaOH and CS_2, the undyed historic cotton fiber displayed patterns of splitting different from the other specimens and the dyed historic cotton fiber showed extensive oxidation while the undyed cotton did not. Single fiber infrared microspectroscopy and microchemical reactions provide clues to internal structural differences that may influence the determination of future conservation treatments; fibers from the same generic class and removed from the same artifact display differing reactivity and structure.

In a companion paper, the results of an infrared study of single historic cotton fibers recovered from a deep ocean shipwreck site were reported. Although these fibers display no alteration in chemistry, the crystallinity indices reflect a significant difference with the undyed historic cotton having the highest index (*1*). Further studies were undertaken to elucidate the differences in biodegradation exhibited by the dyed and undyed fibers from a single garment from that site (*2*). These fibers were treated with 18% NaOH and with a mixture consisting of equal parts of 18% NaOH and CS_2, and examined for evidence of the presence or absence of the primary wall. Differences in microchemical reactivity were noted including differences in patterns of fragmentation in response to the treatments (*2*).

Many approaches have been employed to assess fiber polymer degradation. Examination of fibers under the microscope is a standard method for study of morphology, while infrared spectroscopy provides functional group characteristics.

While solubility is a method sometimes used for estimation of polymer damage (*3*), classic microscopic techniques are often the first step in examination of fibers including observation of microchemical reaction and solubility as a means for identification and characterization (*4*). Alkaline solutions, particularly NaOH, are often employed in the study of natural cellulose fibers; the consequent swelling is due to intracrystalline penetration of the solvated dipole hydrate (*5*). The fibers swell and straighten, resulting in a smoother more lustrous material. If the fiber is cut into fragments, a protrusion can be observed at the cut ends of the fibers under the microscope as the secondary wall swells while the primary wall resists swelling. This is the so-called mushroom or pinhead reaction (*4,6*). It is commonly used for examination of damage to cotton, since, if the primary wall is damaged, it will not resist the swelling of the fiber exposed to an alkaline solution and the characteristic bulges will not be formed. Change in fiber diameter in response to treatment in alkali, then, reflects the condition of the fiber's physical microstructure and can reveal abnormal swelling of biodeteriorated fibers (*7*). Treatment with a mixture consisting of equal parts of 18% NaOH and CS_2 is also employed to detect primary wall damage in cotton (*8-10*). Intact fibers will display the formation of beads or balloons along their length as the primary wall acts to resist swelling while the secondary wall swells upon exposure to the alkali.

When cellulose treated with alkali is exposed to air, oxidative degradation occurs, resulting in chain scission and deterioration of the fiber's properties. Oxidation of alkaline cellulose is an autocatalytic process following a free radical mechanism (*5,11*), characterized by the formation of carbonyl groups. Cardamone et al (*12*) used Fourier Transform Infrared spectroscopy to monitor the formation of carbonyl groups as cellulose oxidized. These authors and others (*13-15*) demonstrated that the FTIR microspectrometer is capable of providing structural information of cotton fibers, and that second derivative spectra can be used to resolve coupled peaks and weak shoulders. In the research reported herein, light microscopy and infrared microspectroscopy were employed to examine the physical and chemical structure of dyed and undyed historic cotton fibers upon treatment with 18% NaOH and with a mixture consisting of equal parts of 18% NaOH and CS_2.

The results of this work not only provide qualitative and quantitative data describing the types and extent of degradation incurred by the treated historic fibers but also reveal their reactivity toward the alkaline treatments. Therefore, these results can provide conservators with additional information about the structure and property differences of the cotton fibers obtained from different fabrics within the same object; these may reflect differing treatment histories of the fibers and also may be indicative of differences in reactivity to potential conservation treatments.

Research Methods

Samples and Sample Treatments

Four samples, described in detail in (*1*) were employed in this work. These consist of a dyed and undyed cotton removed from the back (twill weave) and lining (plain weave) of a man's vest recovered from a shipwreck site labeled MD and MU, a reference cotton (Testfabrics No. 400U), labeled SD, and the same reference cotton placed at the deep ocean site for a period of three months, labeled M3. One set of fiber specimens removed from each of these samples were mounted on slides using Permount (Fisher Scientific) for measurement of diameter of untreated fiber. A second set of individual fiber specimens were each mounted on four to five microscope slides with 18% NaOH and observed under the microscope. After five minutes, the diameter of 50 to 70 fibers was measured from each specimen. Yarn specimens were employed in preparation of the third and fourth sets of fiber specimens due to the washing and drying involved. The third set of fiber specimens was obtained from yarns treated with a mixture consisting of equal parts of 18% NaOH and CS_2 in a test tube for 30 minutes and these fibers were examined microscopically for evidence of the ballooning phenomenon. The treated yarns were exposed to air for 24 hours, collected, washed, and air-dried. Fiber specimens obtained from these yarns were then examined with FTIR microspectroscopy. The fourth set of fiber specimens was obtained from yarns treated with 18% NaOH for 15 minutes, washed and air dried, and then examined with FTIR microspectroscopy.

Instruments and Sample Preparation

A Zeiss Axioplan Research Microscope with differential interference contrast (DIC), phase contrast, and polarized illumination capabilities was used at nominal magnifications of 400X and 1000X. Micrographs of the fibers were taken with a MC 100 35mm camera attached to the Zeiss microscope. Fiber width was measured using a Zeiss Axioskop coupled with an Optimas 5.2 image analysis system at nominal 100X magnification.

The same instrument, instrumentation parameters, and sample preparation method described in the companion paper (*1*) were used for single fiber infrared spectra collection, which is a Bruker Equinox 55 FTIR microspectrometer with a liquid nitrogen cooled narrow band Mercury Cadmium Telluride (MCT) detector. The operation parameters for collecting cotton fiber infrared spectra include 4.0 cm^{-1} resolution and 260 accumulated number of scans. The IR spectra were processed through OPUS IR spectroscopic software (version 2). Second derivative analysis was achieved with Grams 386 (Galactic Industries).

With the exception of the undyed historic marine cotton fibers treated with 18% NaOH, fibers were first slightly flattened by means of a micro-roller device. Each flattened fiber was mounted across a 2mm slit cut into a heavy cardboard frame. Because of their fragility, the undyed fibers and the fiber fragments collected after treatment by a mixture consisting of equal parts of 18% NaOH and CS_2 could not be pressed. Instead they were placed on a 13mm KBr window located on a rectangular infrared compression cell holder. All infrared spectra were collected with the fibers oriented in the same vertical orientation with respect to the microscope.

Results and Discussion

Swelling Behavior

Table I summarizes the results of fiber width measurements before and after 18% NaOH treatment of the four cotton specimens. Before the swelling treatment, the mean fiber width of the dyed historic cotton fiber (MD), 23.6µm, and the undyed historic cotton (MU), 20.3µm, were significantly different from each other, and both were significantly larger than those of the reference cotton (SD) and cotton immersed for 3 months (M3), 16.2µm and 17.3µm respectively, based on a Tukey's pairwise comparison analysis with an alpha level of 0.05. After treatment with 18% NaOH, the fiber width of all four cotton samples increased, but to differing degrees. The percentages of swelling calculated from the fiber width changes are 58.9%, 59.0%, 42.1% and 66.4% for SD, M3, MD and MU respectively.

Table I. Degree of Swelling of the Four Cotton Fibers in NaOH, µm.

Comparison of cotton fiber width before and after NaOH treatment

Sample	Before NaOH treatment			after NaOH treatment			swelling (%)
	Mean	stdev	SE	mean	Stdev	SE	
SD	16.2	2.98	0.183	25.8	4.75	0.184	58.9
M3	17.3	3.15	0.182	27.6	4.28	0.155	59.0
MD	23.6	4.64	0.197	33.3	5.66	0.170	41.1
MU	20.3	3.48	0.171	33.8	6.05	0.179	66.4

SD = Reference cotton, M3 = three-month immersed reference cotton, MD =dyed historic marine cotton, MU = undyed historic marine cotton; SE = standard error (Stdev/mean).

The results presented in Table I indicate that the brown-colored cotton fiber recovered from the marine environment (MD) is significantly larger than the other three specimens. It is possible that it is a different cotton species, perhaps an Indian cotton since its fiber diameter more closely approximates that of Indian cotton (20-23µm) (12, p46). Also this possibility is consistent with the coarse morphological appearance and large lumina of these fibers when viewed under the microscope. Fibers from the undyed historic cotton have diameters which are encompassed by both the Indian cotton and American Upland cotton (16.8-20.2 µm) (12, p46), but do not show the same coarse morphological appearance as those of the dyed cotton. Three months of immersion in the deep ocean produced no significant change in the diameter of the reference cotton and had no effect on the fiber's swelling behavior. One hundred thirty years, however, caused enough alteration of the undyed cotton that it showed a higher degree of swelling than the reference cotton. This high degree of swelling and absence of a mushroom behavior at fiber cut end are indicative of primary wall degradation of the fiber. The MD fibers swell to a much smaller extent than any of the other specimens, reflecting its difference as a different cotton or reflecting the influence of the dye and mordant composition in restricting swelling behavior and reactivity. Further studies by the authors are underway to compare the swelling properties of American Upland cotton and Indian cotton, and to investigate the effect of different dyes used historically on the fiber's swelling properties.

Ballooning Behavior

After 30 minutes of immersion in a mixture consisting of equal parts of 18% NaOH and CS_2, ballooning phenomena were observed in the majority of the reference cotton fibers studied (Figure 1-A). Ballooning of the three-month immersed cotton fibers was not as predominant as in the modern cotton fibers (Figure 1-B). Both the dyed and undyed historic cotton fibers failed to display ballooning behavior, an indication of primary wall damage (Figure 1-C and 1-D).

Fibrillation in the longitudinal direction of the fiber axis was observed in all four samples. However, such decomposition is a predominant feature in the dyed historic cotton fibers (Figure 1-C). Alternatively, many undyed historic cotton fibers displayed a pattern of splitting into fine horizontal elements that presently cannot be explained.

FTIR Spectra of Cotton Fibers Treated with 18% NaOH

The infrared spectra of the four cotton samples treated with 18% NaOH were consistent with the infrared results reported in the literature on cellulose II (Table II). However, treated fibers of the historic undyed cotton displayed spectra that were much more uniform than those from the untreated fibers. It is conjectured that the treatment, e.g. swelling, washing, and drying, biased the ultimate sample of fibers by eliminating the severely degraded fibers.

The spectra presented in Figure 2 demonstrate that the spectra from the four treated cotton specimens are similar and consistent with each other. The similarity and consistency reflect the similar chemical reactivity of the four specimens toward the alkaline reagent, another indication that the chemical structure of the cotton has not been altered by the marine environment.

The OH stretching region

A peak located at 3488 cm^{-1} (Figure 3) was revealed by the second derivative analysis of the NaOH treated cotton samples as that identified by polarized IR (*16*). For most mature cotton fibers however, fewer peaks were resolved by second

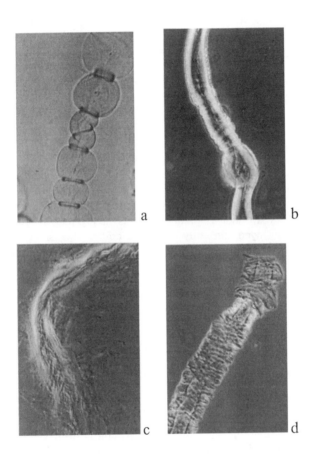

a = reference cotton   b = three-month immersed cotton
c = dyed historic cotton   d = undyed historic cotton
Figure 1. Light microscope micrographs of cotton fibers after treatment with a ballooning agent (400x).

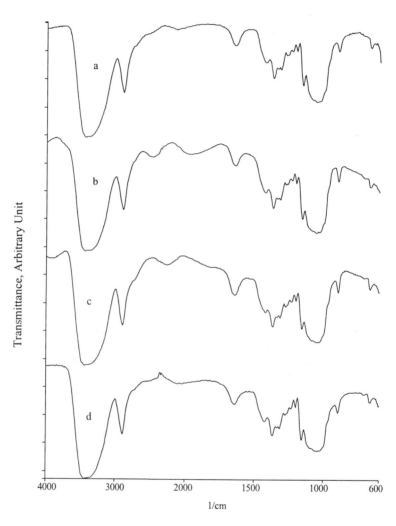

a = reference cotton   b = three-month immersed cotton
c = dyed historic cotton   d = undyed historic cotton

Figure 2. Infrared spectra of cotton fibers after 18% NaOH. treatment.

Table II. Infrared Peak Identified from Cotton Samples Treated with 18% NaOH.

Peak identification from normal IR spectra and 2nd derivative calculation								Results from Literature	
SD		M3		MD		MU		Cellulose II	
N	2nd D	N	2nd D	N	2nd D	N	2nd D	P. IR (16)	2nd D (17)
	3542								3553
	3534								3529
	3500				3500		3499		3505
	3488		3489		3488		3488	3488	
					3473		3473		3474
3444	3443	3445	3444	3442	3442	3443	3443	3448	3442
	3415				3417		3417		
					3406		3406		3404
								3350	3361
	3341								3330
									3313
								3305	
								3175	
	2983		2983		2983		2983	2981	2984
	2969		2969		2970		2969	2968	2965
	2956		2956		2957			2955	2956
							2944		
	2935		2935		2935		2934	2933	2935
	2919		2919		2918		2918		2922
	2905		2905		2906		2906	2904	2903
2893	2891	2895	2891	2894	2892	2895	2892	2891	2892
	2875		2874		2875		2972	2874	2875
	2850		2850		2850		2849	2850	2849
	1470		1470		1470			1470	
			1463		1463		1463		
			1454				1454		
			1444					1440	
			1434				1432		
1422	1419	1423	1422	1418		1422	1421		
			1415	1415					1416
1369	1376	1370	1375	1368	1377	1368	1373	1375	
	1363		1364		1365		1363	1365	
1336	1335	1336	1335	1337	1337	1336	1337	1335	
1315	1314	1316	1315	1315	1315	1316	1316	1315	
1277	1277	1278	1278	1277	1277	1277	1279	1277	
1264	1261	1262	1262	1264	1263	1264	1263	1257	
1232	1234	1232	1234	1234	1234	1234	1235		
	1225		1226	1227	1226			1225	1225
1200	1200	1200	1200	1200	1199	1200	1200	1200	

Table II. Infrared Peak Identified from Cotton Samples Treated with 18% NaOH.

Peak identification from normal IR spectra and 2nd derivative calculation								Results from Literature	
SD		M3		MD		MU		Cellulose II	
N	2nd D	N	2nd D	N	2nd D	N	2nd D	P. IR (16)	2nd D (17)
	1166		1162		1161		1164		
1157	1158	1157		1157		1157		1155	
			1123		1123		1123		
			1112		1112		1111	1107	
								1078	
1059		1064	1059			1059	1060	1060	
1034					1034		1034	1035	
			1019		1018		1021	1020	
								1005	
	989		996		994		995	996	
	962		964		963		963	965	
896	895	896	896	896	896	896	896	892	
								800	
								760	
703		701					712	700	
								650	

SD = Reference cotton, M3 = three-month immersed reference cotton, MD =dyed historic marine cotton, MU = undyed historic marine cotton; N = Normal infrared spectra, 2nd IR = Second derivative infrared spectra, P. IR = Polarized infrared.

derivative calculation in this study than the number of peaks reported in the second derivative study by Michell (*17*). This may be due to the different sample preparation methods employed between the two studies, one using a KBr pellet and the other using an oriented single fiber. In a deconvolution study of the OH region of cotton fibers treated with various concentrations of NaOH, more peaks were resolved from cotton treated with higher concentrations of NaOH than cotton treated with lower concentrations of the alkali (*18*). The variation of peak positions reported in these different studies may be due to variation in sample preparation between the studies, since both concentration and temperature can affect the degree of swelling of the cotton fiber.

The CH stretching region

The CH/CH_2 stretching peak in the infrared spectra of NaOH treated cotton samples appeared different from that of the untreated cottons. First, the position of the CH/CH_2 stretching peak of the treated cotton is at 2894 cm^{-1} instead at 2900 cm^{-1} in the spectra of untreated cotton. Second, the shoulders around the peak at

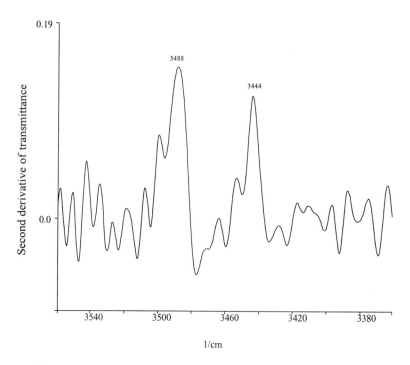

Figure 3. Second derivative infrared spectrum of OH stretching region of reference cotton after 18% NaOH treatment.

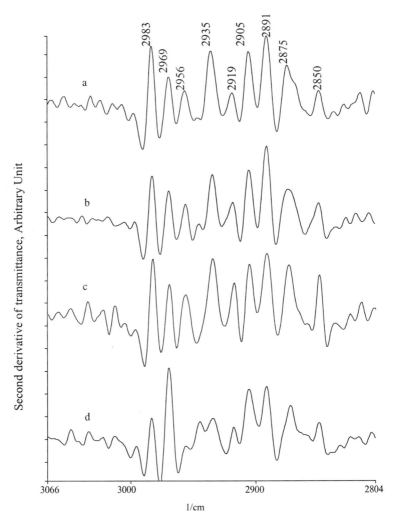

a = reference cotton   b = three-month immersed cotton
c = dyed historic cotton   d = undyed historic cotton

Figure 4. Second derivative infrared spectra of cotton after 18% NaOH treatment.

2894 cm^{-1} of the treated cottons are not as distinct as those around the peak at 2900 cm^{-1} in the untreated cottons (Figure 2). The spectra of the fibers examined in this work appear similar to the spectra of mercerized cottons reported in the literature. All of the peaks identified in the polarized IR study by Marchessault and Liang (16) and in the second derivative study by Michell (17) were also identified in all of the four NaOH treated cotton specimens examined in this study (Figure 4).

The strongest peak in cellulose II, at 2894 cm^{-1}, is assigned as CH stretching by Liang (19). The infrared absorption due to CH$_2$ stretching is assigned to the peaks at 2933 cm^{-1} and 2850 cm^{-1} in cellulose II. Liang (19) suggested that CH$_2$OH was not involved in the intramolecular hydrogen bonding in cellulose I and II. However, the new peak that appeared in cellulose II at 2981cm^{-1} in the literature and also in the spectra obtained in this research was not explained. Blackwell (20) proposed a different structure in which CH$_2$OH is involved intramolecular hydrogen bonding in cellulose I and II. In this study, the second derivative spectrum of the undyed historic cotton fiber showed some variations in the resolved peak pattern from those of the other fibers, though the peak positions are similar to the others (Figure 4). This may be attributed to the higher degree of swelling that occurred in the undyed cotton fiber.

The region from 1500 cm^{-1} to 1200 cm^{-1}

The infrared peaks in this region are assigned to the deformation modes of CH and OH, similar to those of cellulose I. In comparison to the same region of cellulose I, changes in peak position and relative intensity were observed in cellulose II (Figure 2, Figure 5). The peak observed at 1429 cm^{-1} assigned as CH$_2$ symmetric bending in cellulose I shifts to 1420 cm^{-1} in cellulose II (21). The relative intensity of this peak decreased significantly upon the transformation from cellulose I to cellulose II. Both of these features were observed in spectra of the four treated cotton samples in this research. As mentioned in the literature and also observed in this research, one of the OH in-plane bending modes at 1335 cm^{-1} from cellulose II is at the same position as that of cellulose I, but the other two OH bending modes at 1455 cm^{-1} and 1205 cm^{-1} from cellulose I shift to 1470 cm^{-1} and 1200 cm^{-1} in cellulose II. The second derivative spectrum of undyed historic cotton in this region also showed some variations in the resolved peak pattern from those of the others, especially around 1450 cm^{-1} (Figure 5). These variations of the undyed cotton can be related to the variations showed in the CH/CH2 stretching region observed in figure 4, which is attributed to the higher degree of swelling that occurred in the undyed historic cotton fiber.

The region from 1200 cm^{-1} to 600 cm^{-1}

With the exception of the peak observed at 1157 cm^{-1}, assigned as anti-symmetric bridge COC stretching by Liang (19), the other peaks are all coupled into a broad peak (Figure 2). The bands of this region are assigned to the vibration modes relating to CO, or COC, or ring stretching. The increased coupling of the peaks in this region, like that of the CH stretching region, indicate the increased complexity of the hydrogen bonding network in cellulose II. Another characteristic observed in NaOH treated cotton samples in this study is the increased peak intensity of the peak around 895 cm^{-1} in comparison with that of the untreated cotton (Figure 2).

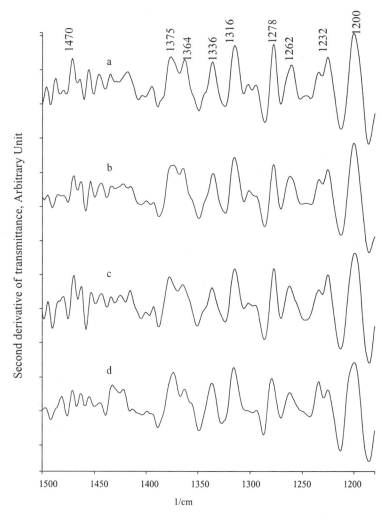

a = reference cotton   b = three-month immersed cotton

c = dyed historic cotton   d = undyed historic cotton

Figure 5. Second derivative infrared spectra of cotton after 18% NaOH treatment.

FTIR Spectra of Cotton Fibers Treated with ballooning agent

The undissolved cotton fibers collected after 30 minutes of treatment with a mixture consisting of equal parts of 18% NaOH and CS_2 were found to display the same infrared spectral characteristics as those treated with NaOH alone. After a 24 hour exposure to air, however, fiber fragments from the dyed and undyed historic cotton fiber displayed different infrared spectra due to the oxidation and severe fibrillation that occurred.

The spectra of dyed historic cotton fibers (MD) after this prolonged strong swelling treatment showed obvious variations between the spectra from different fragments (Figure 6-A), due to the heterogeneous reaction of the fibers with the solution. The common features observed in these spectra are the peak at 1766 cm^{-1} and also peaks around 2500 cm^{-1}. The second derivative calculation resolved a pair of peaks at 1768 and 1754 cm^{-1} (Figure 6-B) indicating oxidative formation of C=O bonds in the cellulose molecules.

In order to confirm that these two peaks are due to oxidized cellulose and not to residual alkaline reagent, infrared spectra of the NaOH and CS_2 reagents reported in literature were studied. The NaOH spectrum has a very weak peak at 1775 cm^{-1}, while the carbonyl peak in the treated cellulose is relatively strong, and the peak position is also different (*17*). The infrared peaks of CS_2 are located at 2300 cm^{-1}, 2200-2100 cm^{-1}, and 1600-1400 cm^{-1}. Thus, the reagents themselves provide peaks that are exclusive of oxidized cellulose and the peaks observed in the treated fiber spectra are therefore ones due to fiber oxidation.

The peaks around 2500 cm^{-1} observed in the oxidized fibers may be due to residual NaOH in the fiber fragment. In examination of the NaOH spectrum (*17*), peaks at 2496 cm^{-1} and 2573 cm^{-1} were observed along with a stronger peak at 879 cm^{-1} and a large broad peak around 1416 cm^{-1}. However, in an earlier study of highly oxidized monocarboxycellulose, a weak band at 2500 cm^{-1} was observed that was interpreted as perturbed carboxylic hydroxyl groups involved in hydrogen bonding (*22*).

The spectra of undyed fiber (MU) after the same prolonged strong swelling treatment showed different characteristics and did not display carbonyl peaks or oxidative features (Figure 7-A and 7-B). The fiber fragments that remained after the prolonged strong swelling treatment were fibrillated. Only two or three peaks were identified in the CH/CH$_2$ stretching region, 2961cm^{-1}, 2922 cm^{-1}, and 2853 cm^{-1}. When the spectra were acquired from loose, partially dissolved fibrillar fragments, the CH/CH$_2$ stretching vibration was very weak, and only the CH$_2$ stretching peaks could be identified (2853 and 2922 cm^{-1}). The intensity of the CH/CH$_2$ peaks is proportional to the cellulose content in the IR radiation path and the difference in peaks in this area could reflect the disruption of molecular order and association of the severely degraded and fibrillated undyed cotton fiber. The differences in response to the treatment cannot be completely explained but they do reflect a difference in susceptibility between the dyed and undyed historic cotton fibers. Additional peaks observed include a water band around 1635 cm^{-1}, and a few bands related to C-O and ring vibrations at 1200 cm^{-1}, 1162 cm^{-1}, 1124 cm^{-1}, 1029 cm^{-1}, and 999 cm^{-1}.

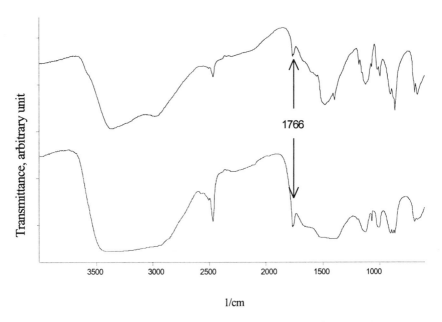

Figure 6-A. Infrared spectra of dyed historic cotton fibers after NaOH + CS₂ treatment.

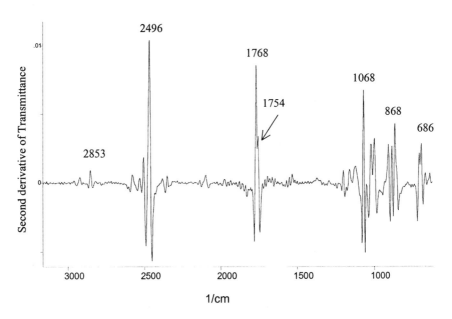

Figure 6-B. Second derivative spectrum of dyed historic cotton after NaOH + CS₂ treatment.

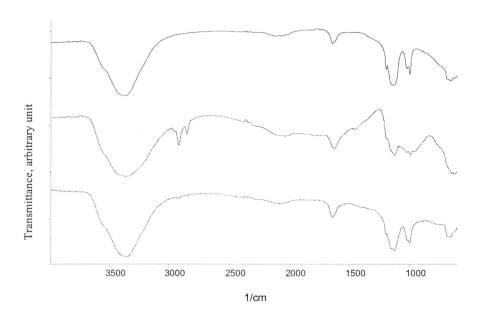

Figure 7-A. Infrared spectra of undyed historic cotton after NaOH + CS₂ treatment.

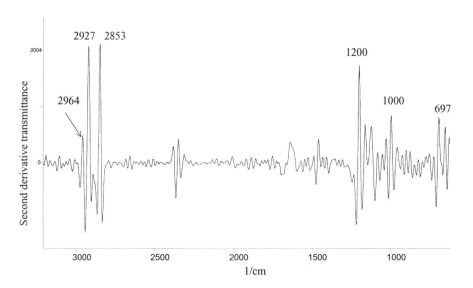

Figure 7-B. Second derivative spectrum of undyed historic cotton after NaOH + CS₂ treatment.

The difference in reactivity of the dyed versus the undyed historic cotton fiber is due to the tin it contained in the dyed fiber that had been used as a mordant in the dyeing process (23). Since studies have shown that certain metals, such as Fe(II), Co(II), Mn(II), Cu(II), had positive catalyzing effects towards the oxidation of alkaline cellulose (24-27), it can be conjectured that the tin catalyzed the oxidation of the dyed historic cellulose when exposed to a mixture consisting of equal parts of 18% NaOH and CS_2. In this study, no infrared bands attributable to dye components in the dyed cotton fiber were observed. It is possible that the dye concentration in the fibers is so low that it is beneath the detection limit of the method or the absorbance bands are very small in comparison to those of the fiber (28). On the other hand, our examination of infrared spectra of some natural brown dyes has indicated that the chemical constituents are similar to those of cellulose and produce weak IR bands that could be masked by the IR bands of cotton. Future studies need to address identification of dyes in these historic marine textiles.

Conclusions

Treated with a mixture consisting of equal parts of 18% NaOH and CS_2, both undyed and dyed historic marine cotton fibers did not show ballooning phenomena, indicating that both specimens were biodegraded; the primary wall was damaged and could not restrain swelling of the secondary wall. The dyed historic cotton fibers are dominated by the fibrillation along the fiber axis direction, and the undyed cotton fibers are dominated by horizontal fragmentation or splitting across the fiber. With 18% NaOH treatment, the undyed fibers showed a much higher degree of swelling than any of the specimens while the dyed fibers showed a much smaller degree of swelling. Both the change in fiber width after 18% NaOH treatment and the pattern of swelling and dissolution in a mixture of NaOH and CS_2 demonstrate that a higher degree of biodegradation occurred in the undyed historic cotton, and that the dyed and undyed fibers reacted differently with the same chemical treatments.

Comparison of infrared spectra and second derivative spectra of the cotton fibers treated with NaOH indicated that the structural transition from cellulose I to cellulose II was similar to those reported in the literature. Due to the higher degree of swelling that occurred in the undyed historic cotton, the second derivative spectrum showed some variations in the resolved peak pattern from those of the other cotton fibers. However, the infrared peak positions identified from the four NaOH treated cotton samples were similar. In addition to the results obtained in the FTIR study of untreated cotton fiber specimens (1), the behavior of the treated fibers and the consequent infrared spectra provide a further confirmation that the cellulosic chemical structure of the historic cotton fibers was not altered even after 133 years exposure to the deep-ocean environment. The infrared study also demonstrated that the difference in spectra between cellulose I and cellulose II can be easily detected in the CH/CH_2 stretching region via calculation of second derivative spectra. The results of FTIR analysis of samples treated with 18% NaOH indicate that the two historic cotton fibers react with the alkaline solution in a chemically similar manner, yet the physical consequences of this treatment are dissimilar.

Comparison of the infrared and second derivative spectra collected from fibers after prolonged treatment in a mixture consisting of equal parts of 18% NaOH and CS_2, indicate that the dyed historic cotton was oxidized while the undyed historic cotton was not. The different reactions between the two materials are likely due to the tin mordant present in the dyed cotton that would serve to catalyze oxidation.

Although similar in chemistry, i.e., both are cellulose, the physical and chemical microstructure of the dyed and undyed historic cotton fibers obtained from the same textile artifact was shown to be different as a result of the differences they display in reaction to treatment with alkaline solutions. Therefore, any future conservation treatment of the artifact should include consideration of these differences. The results of this study of two cotton fabrics incorporated in a single garment also show that before the development and application of any conservation treatment, studies on each of the components of the object are necessary.

For example, although agents employed for wet cleaning textiles are formulated to establish a neutral pH, and the reagents used in this work are alkaline, the differing reactivity of the fibers could result in differing results in conditions typical of wet cleaning. Microchemical reactivity and single fiber infrared microspectroscopy are tools by which information can be gained from a very small sample concerning the condition of a textile artifact. The reaction of individual fibers to proposed cleaning agents could be observed under the microscope and the treated fibers then examined with infrared microspectroscopy to determine how a treatment will affect an entire garment. These observations may also prove useful in inferring past treatments of the textile as well.

References

1. Chen, R.; Jakes, K. A. In *Historic Textiles, Paper and Polymers in Museums*, J. Cardamone, Ed.; ACS symposium series, American Chemical Society: Washington, D.C., accepted for publication.
2. Chen, R.; Jakes, K. A. Postprints of the Textile Specialty Group. American Institute for Conservation, accepted for publication.
3. Crighton, J. S. In *Polymers in Conservation*, Allen, N. S.; Edge, M.; Horie, C. V. Eds. Royal Society of Chemistry: Cambridge, 1992; pp.82-107.
4. Heyn, A. N. J. *Fiber Microscopy*; Interscience Publishers, Inc.: New York, 1954.
5. Freytag, R.; Donze, J. J. In *Chemical Processing of Fibers and Fabrics: Fundamentals and Preparation* Part A, Lewin, M.; Sello, S. B. Eds.; Marcel Dekker, Inc.: New York, 1983; pp.93-166.
6. Mahall, K. *Quality Assessment of Textiles: Damage Detection by Microscopy*; Springer-Verlag: Berlin, 1993.
7. Heyn, A. N. J. *Textile Ind.* **1956**, *120*, 137-145.
8. Thaysen, A. C.; Bunder, H. J. *J. Roy. Microscop. Soc.* **1923**, *21*, 303.
9. Siu, R. G. H. In *Cellulose and Cellulose Derivatives*, Ott, E.; Spurtin, H. M.; Grafflin, M. W. Eds.; Interscience: New York, 1954; pp.183-191.

10. Allen, S. J.; Auer, P. D.; Pailthorpe, M. T. *Textile Res. J.* **1995**, *65*(7), 379-385.
11. Nevell, T. P. In *Cellulose Chemistry and Its Applications*, Nevell, T. P.; Zeronian, S. H. Eds.; Ellis Horwood Limited: England, 1985; pp. 243-265.
12. Cardamone, J. M.; Keister, K. M.; Osareh, A. H. In *Polymers in Conservation*, Allen, N. S.; Edge, M.; Horie, C. V. Eds. Royal Society of Chemistry: Cambridge, 1992; pp.108-124.
13. Chen, R. Ph.D thesis, The Ohio State University, Columbus, OH 1998.
14. Michell, A. J. *Carbohydrate Res.* **1993**, *241*, 47-54.
15. Pandey, S. N. *Textile Res. J.* **1989**, *59*, 226-231.
16. Marchessault, R. H.; Liang, C. Y. *J. Polym. Sci.* **1960**, *43*, 71-84.
17. Fengel, D.; Strobel, C. *Acta Polym.* **1994**, *45*, 319-324.
18. Michell, A. J. *Carbohydrate Res.* **1988**, *173*, 185-195.
19. Liang, C.Y.; Marchessault, R. H. *J. Polym. Sci.* **1959**, *37*, 269.
20. Blackwell, J. In *Cellulose Chemistry and Technology*, J. C. Arthur, Jr. Ed.; ACS Symposium Series, 48, American Chemical Society: Washington, DC., 1977; pp.81-86.
21. Sao, K. P.; Mathew, M. D.; Ray, P. K. *Textile Res. J.* **1987**, *57*, 407-414.
22. Zhbankov, R. G. *Infrared Spectra of Cellulose and Its Derivatives*; Consultants Bureau: New York, 1966.
23. Crooks, W. *Dyeing and Calico-Printing*, Longmans, Green, and Co.: London, 1874.
24. Warwicker, J. O. In *Cellulose and Cellulose Derivatives* Part IV, Bikales, N. M.; Segal, L. Eds.; Wiley-Interscience: New York, 1971; pp. 325-380.
25. Nevell, T. P. In *Cellulose Chemistry and Its Applications*, Nevell, T.P.; Zeronian, S. H. Eds.; Ellis Horwood Limited : England, 1985; pp. 243-265.
26. Nevell, T. P.; Singh, O. P. *Textile Res. J.* **1986**, *56*, 271.
27. Kokot, S.; Marahusin, L.; Schweinsberg, D. P. *Textile Res. J.* **1994**, *64*, 710-716.
28. Katon, J.E. *Micron* **1996**, *5*, 303-314.

Chapter 6

Degradation and Color Fading of Cotton Fabrics Dyed with Natural Dyes and Mordants

N. Kohara[1], C. Sano[2], H. Ikuno[3], Y. Magoshi[4], M. A. Becker[2,5], M. Yatagai[6], and M. Saito[7]

[1]Faculty of Practical Arts and Science, Showa Women's University 1–7 Taishido, Setagaya-Ku, Tokyo 154–8533, Japan
[2]Conservation Science Department, Tokyo National Research Institute of Cultural Properties, 13–27, Ueno Park, Taito-Ku, Tokyo 110–8713, Japan
[3]Department of Clothing Science, Tokyo Gakugei University, 4–1–1, Nukuikita, Koganei, Tokyo 184–8501, Japan
[4]NISES, Tsukuba 305–8634, Japan
[6]Tokyo Gakuen Women's College, Tokyo 158–8586, Japan
[7]Kyoritsu Women's University, Tokyo 101–8433, Japan

The effects of six mordants, iron and aluminum salts, and two major components of natural dyes, hematoxylin and curcumin, on the photodegradation rates of cotton fabrics were evaluated. Samples were mordanted or dyed-mordanted according to traditional Japanese methods and then exposed to simulated sunlight for incremental doses up to 500 hours. Changes in color and the tensile properties were evaluated. There was a dramatic color change in the dyed and alum-mordanted samples compared with the dyed and iron sulfate-mordanted fabrics. However, the iron salts-mordanted fabrics exhibited a greater loss in tensile properties after 500 hours. For the dyed and alum-mordanted fabrics, there appeared to be a preferential degradation of the dye to the fiber.

Ancient people in different regions had their own techniques for making textiles within their own culture. They used different fibers such as linen in Egypt, cotton in India, silk in China and wool in Central Asia. Before the discovery of synthetic dyes,

[5]Current address: Fukui University, Fukui-shi 910–8507, Japan.
[6]Current address: Chiba University, Chiba-shi 263–8522, Japan.
[7]Corresponding author.

textiles had been colored with natural dyes for several thousand years. Most of the natural dyes were dyed with mordants, mainly aluminum, iron and tin salts. But in Japan, tin was not often used. Even today, some textiles in Japan are still mordanted in mud to fix some natural dyes. It is well known that these metal salts play an important role in the color fading of the dye and the degradation of the fiber substrate. Rates of degradation of historic textiles differ depending on types of fibers, dyes, mordants and the combination of them. For example, textiles mordanted with aluminum salts sometimes show considerable color-fading, while iron compounds are often regarded as one of the substances which can accelerate the degradation of fibers (1). Numerous research papers on the degradation and color fading of natural fibers dyed with synthetic dyes have appeared in the literature (2-7). However, the quantitative influence of mordants and natural dyes on degradation of natural fibers and the mechanism had not been clarified enough in spite of importance to the conservation of historic textiles.

The objective of our study is to gain a better understanding of the deterioration of natural-dyed fabrics, including the degradation of the fiber as well as the decomposition of dye molecules on the fiber. Natural dyed fabrics of cellulosic fibers (cotton and linen) and protein fibers (silk) have been prepared using several combinations of dyes and mordants, and the effects of dye, mordant and fiber substrate on the deterioration of the fabrics caused by light have been investigated by means of various analytical techniques. The photo-degradation of linen dyed with curcumin (CUR) or hematoxylin (HEMA) and mordanted with aluminum or iron salts similarly to the cotton fabrics was previously reported (8). It was found that the linen dyed with CUR and mordanted with alum showed the most rapid color fading in all of the dyed and mordanted linens but showed the least loss in tensile strength. The experimental results suggested that preferential degradation of dye provides protection to the fiber from light degradation. This paper presents the results of the studies on cotton and the results of the silk studies are described elsewhere (9). In this work the effects of several factors, including dye, mordant and absence of impurities in cotton, on the degradation of cotton fiber assessed by strength loss and color change have been studied.

Experimental

Materials

A grey cotton fabric with a plain-weave construction (50 × 30 yarns/cm, 144 g/m^2) purchased from Shikisen-sha Co. Ltd. was the test fabric. This fabric was desized by boiling in a 0.2% nonionic surfactant aqueous solution for 20 min and then washed with distilled water (the desized grey material: **G**). Part of the grey material was scoured-bleached by boiling in an aqueous solution containing 0.8%v/v H_2O_2 (35% conc.), 0.7%v/v NaOH and 0.05%v/v nonionic surfactant for 20 min (liquor ratio, 17:1), followed by neutralization with formic acid and thorough washing with distilled water (the scoured and bleached material: **S**). Cotton counts of the warp yarns from grey and the scoured and bleached fabrics were 3.9 and 4.2, respectively. Purity grade and sources of the dyes and metal salts used are as follows: Curcumin (**CUR**, CI 75300) and hematoxylin (**HEMA**, CI 75290) purchased from Merck Co., which were respectively 97% and >95% pure. Reagent grade aluminum potassium sulfate (**AL1**), aluminum chloride (**AL3**), iron (II) sulfate (**FE1**) and iron (II) chloride (**FE3**) were purchased from Wako Pure Chemicals Industries Ltd. Aluminum acetate containing Al_2O_3 at 13~14% from Terada-yakusen Industry Co. (**AL2**) and iron (II) acetate (**FE2**) was 95% pure from Aldrich Chemical Co.

Dyeing and Mordanting Procedure

Fabric samples, 15 × 10 cm (grey fabric: 2.2 g, scoured and bleached fabric: 2.1 g) were weighed and mordanted and/or dyed as follows. Dyeing and mordanting processes were carried out according to Japanese traditional methods (*10*).

Mordanting
Aluminum and iron mordant solutions respectively contained 8% aluminum or 14% iron on weight of fabric (owf) and a liquor ratio of 30:1 was used. The sample fabrics were placed into the mordant solutions at 21°C for 30 min with constant agitation. Mordanted samples were then thoroughly rinsed with distilled water and allowed to air dry. All undyed and mordanted samples were prepared only by once mordanting with no dyeing. Undyed-mordanted fabrics are represented by the abbreviations of the used fabrics and mordants such as **G-AL1, S-FE1** and so on.

Dyeing
The dye baths contained 3% curcumin or hematoxylin owf, the pH was adjusted to 7.8 or 8.8~9.0, respectively by adding sodium carbonate and a liquor ratio of 30:1 was used. The fabrics were immersed in the dye bath at 21°C, and the bath was brought to 80~85°C and maintained at this temperature for 30 min with constant agitation. The dyed fabrics were then mordanted respectively with alum or iron sulfate. For the dyed and mordanted samples, the preparation work was accomplished by alternatively dyeing (three times) and mordanting (two times). After the treatment

was completed, the dyed-mordanted samples were rinsed thoroughly with distilled water and were allowed to air dry. Dyed-mordanted samples were represented by the abbreviations of the used fabrics, dyes and mordants, for example **GCUR-AL1** and **SHEMA-FE1**.

Weathering Tests

Undyed-mordanted and dyed-mordanted cotton fabrics were exposed to simulated sunlight in a xenon long-life fade meter, model SUGA FAL-25AX-HC, using simulated sunlight from a xenon source with borosilicate inner and outer filter operating at an average black panel temperature of 63°C and at an average relative humidity of 25% according to JIS L 0803. The two borosilicate filters cut off the light of the wavelength below 275 nm. A xenon lamp, irradiation power of 310 $\pm$ 10 W/m^2 in the range of 300~700 nm, was used for the light exposure of 100 h, 400 h and 500 h. The average incident light energy of the exposure for 100h was 648 MJ/m^2 in the range of 300~3000 nm, which corresponds to the average incident sunlight energy over ca. 55 days in Tokyo. For the purpose of an even exposure, one side of the sample was continuously exposed for certain time and then another side was exposed again for the same period. The total exposure time of both sides of the samples was 500 h at the longest.

Analysis and Measurements

Inductively Coupled Plasma Atomic Emission Spectrometer (ICP-AES) and ESR

Amounts of aluminum and iron in undyed-mordanted or dyed-mordanted samples were measured with using an ICP-AES, model SEIKO SPS-1500VR, at a frequency of 27.12 MHz and Rh power of 1.3 kW, where coolant, auxiliary and carrier gas (Ar) flow rates were respectively 16.0, 0.4 and 0.4 L/min.

ESR spectroscopy of unexposed samples was carried out at 25°C, using a JES-RE2X esr spectrometer operated in the X-band with 100 kHz modulation. **g** Values and intensity were measured by comparison with Mn^{2+} in ZnS and 4-hydroxy-tempo as standard markers, respectively.

Color Measurement

The color expressed as a*, b* and L* color coordinates and color differences (ΔE^*_{ab}) between exposed and unexposed samples were calculated using the CIE 1976 L*a*b* formula as follows:

$L^* = 25(100Y/Y_0)^{1/3} - 16,$

$a^* = 500[(X/X_0)^{1/3} - (Y/Y_0)^{1/3}],$

$b^* = 200[(Y/Y_0)^{1/3} - (Z/Z_0)^{1/3}],$ and

$\Delta E^*_{ab} = [(\Delta L^*)^2 + (\Delta a^*)^2 + (\Delta b^*)^2]^{1/2},$

where X, Y, and Z are the tristimulus values for the specimen and X_0, Y_0 and Z_0 define the color of the normally white object-color stimulus. ΔL^*, Δa^* and Δb^* values were respectively determined as difference between the values of unexposed in this work.

All untreated, undyed-mordanted and dyed-mordanted samples were measured with a MacBeth CE-7000 color spectrometer and illuminant C.

Tensile Tests

Tensile properties including breaking strength, percent elongation at break and energy to break were measured for 20 warp yarns from each sample with an Instron model tensile testing machine, model TENSILON UCT-1500 by using a 20 mm gauge length and a cross head speed of 20 mm/min.

Results and Discussion

Dyeing and Mordanting

Desized grey (G) and the scoured and bleached fabric (S) were mordanted and/or dyed under the same conditions and then exposed to light in order to observe any influence that impurities in the grey fabric had on the deterioration by light. Both kinds of fabrics were only mordanted with six types of aluminum or iron salts (undyed-mordanted) or dyed with CUR and HEMA and mordanted with alum and iron sulfate (dyed-mordanted). The color of the fabrics scarcely changed after only mordanting with aluminum salts, while it slightly changed to a light brownish when mordanted with iron salts (Table I). The combination of dyes and mordants on the cotton fabrics produced the different colors.

From the ICP-AES data shown in Table II, no significant difference is observed between amount of the metal on the undyed-mordanted and on the dyed-mordanted samples. Both aluminum and iron were found to have a higher concentration in undyed-mordanted and dyed-mordanted G samples compared with the corresponding S samples. This suggests that the mordants easily bind to the impurities in grey such as protein, pectin, fat and so on. The amounts of iron were generally more than the amounts of aluminum, because the higher concentration of Fe-mordant solution was used according to Japanese traditional recipe (*10*).

Table I. Color Changes in Undyed-Mordanted and Dyed-Mordanted Cotton Fabrics after Light Exposure for 500 h.

Sample	Unexposed			Exposed for 500 h			
	L^*	a^*	b^*	L^*	a^*	b^*	ΔE^*_{ab}
G	90.1	0.5	8.7	91.3	0.1	6.1	2.9
G-AL1	90.1	0.4	9.1	90.9	0.2	5.4	3.8
G-AL2	90.1	0.5	9.3	90.8	0.3	5.2	4.2
G-AL3	87.7	0.4	5.1	87.7	0.4	5.1	4.8
G-FE1	81.6	3.9	20.8	78.4	4.5	20.6	3.3
G-FE2	77.8	6.7	22.6	76.8	6.7	22.7	1.0
G-FE3	80.4	4.7	21.2	78.0	5.4	21.5	2.5
GCUR-AL1	67.3	20.7	78.4	83.5	2.9	11.8	70.9
GCUR-FE1	30.3	12.4	20.8	38.3	8.0	18.7	9.3
GHEMA-AL1	31.3	7.8	-13.7	68.5	3.1	3.7	41.3
GHEMA-FE1	27.5	1.1	-3.1	34.4	0.6	5.0	10.6
S	93.9	-0.4	3.6	89.6	0.0	3.5	4.2
S-AL1	94.5	-0.5	3.1	92.1	-0.2	3.1	2.4
S-AL2	94.5	-1.7	3.1	91.6	-0.2	3.3	3.3
S-AL3	94.4	-0.5	3.3	89.1	0.0	1.6	5.6
S-FE1	85.5	4.3	21.4	83.5	4.3	20.8	2.1
S-FE2	84.8	4.7	18.4	82.2	5.2	20.0	3.1
S-FE3	85.7	4.0	19.6	83.1	4.8	20.1	2.8
SCUR-AL1	70.9	17.5	77.7	84.0	2.9	11.8	68.7
SCUR-FE1	40.0	14.5	32.5	53.5	5.9	18.3	21.4
SHEMA-AL1	31.7	7.0	-11.8	60.5	3.5	2.2	32.2
SHEMA-FE1	27.8	0.6	-0.9	35.8	0.3	3.7	9.2

Refer to the AATCC Test Method 153-1985, Section 9.2.9 CIE 1976L*a*b*Color difference formula where L* is lightness index difference (higher values are lighter), a* is chromaticity difference (within a range -a to +a green to red, respectively); and b* is chromaticity difference (within a range of -b to +b, blue to yellow, respectively). ΔE^*_{ab} means difference in color between the sample exposed for 500 h and the corresponding unexposed sample.

Hon reported the effect of ferric ions on formation of free radicals in photo-irradiated cellulose (11). If the stable organic radicals as well as the substances which

Table II. Amounts of Al and Fe in Undyed-Mordanted and Dyed-Mordanted Samples as analyzed with ICP-AES.

Sample	Al(mg/g)	Fe(mg/g)	Sample	Al(mg/g)	Fe(mg/g)
G	0.1	0.2	S	0.1	0.1
G-AL1	0.9		S-AL1	0.7	
GCUR-AL1	1.1		SCUR-AL1	0.4	
GHEMA-AL1	0.7		SHEMA-AL1	0.5	
G-FE1		1.8	S-FE1		1.1
GCUR-FE1		1.8	SCUR-FE1		0.9
GHEMA-FE1		1.4	SHEMA-FE1		1.3

can act as a photosensitizer or initiator in photochemical reaction are contained in the fabrics or formed through the treatments before light exposure, these species will accelerate the photo-degradation of the fibers. Unexposed samples were analyzed by esr to find the stable organic radicals latent in the samples which may act as an accelerator in photo-degradation. A broad single peak with the half-height width of 9~10 mT at the **g** value of 2.0040~2.0045 was commonly observed on the spectra of all dyed-Al-mordanted samples. This peak was not observed on the spectra of undyed-Al-mordanted samples nor undyed- and dyed-Fe-mordanted samples. For other mordanted samples prepared similarly to the dyeing and mordanting processes using iron sulfate or alum but without dye, a radical was not detected in these samples. These facts suggest that the radicals result from the interaction between the dyes and the alum. The unexpected fact that the organic radical was not detected in dyed-Fe-mordanted samples cannot be explained from the obtained data and remains to be clarified.

Color Change

Light-induced color change in undyed-Al-mordanted samples was barely detectable, while the light brown color of undyed-Fe-mordanted samples became visually a little deeper after the exposure. The ΔE^*_{ab} values between unexposed and exposed undyed-mordanted samples were almost under 5 units even after the longest exposure (Table I). Influence by the anions of metal salts on color change was less than that by the metals.

The colors of dyed-mordanted samples exposed to light faded at various rates (Table I), which indicates that the type of mordant and dye significantly influenced on the color-fading of the dyed-mordanted samples. The colors of the corresponding dyed-mordanted G and S samples faded in similar way. In both cases of CUR and HEMA, mordanting with alum caused them to undergo greater color change upon light exposure than mordanting with iron sulfate. Samples dyed with CUR and mordanted with alum underwent the largest color change. ΔE^*_{ab} values of these samples were ca. 30 units after 40 h exposure and beyond 60 units after 400 h exposure. At the end of exposure, a* and b* values of the samples respectively dropped and the colors were visually observed to fade almost completely. Color changes in the samples dyed with HEMA and mordanted with iron sulfate had ΔE^*_{ab} value of 9.7 in average at 500 h exposure and were not as significant as those in the samples dyed with CUR and mordanted with alum. The photo-degradation of the linen dyed and mordanted similarly to the cotton fabrics in this study but exposed under slightly different conditions, where only one side of the samples was exposed to filtered xenon light for 480 h, has been reported (9). The order of color-fading rates in the dyed and mordanted linens was similar to the order of the rates in the cotton samples. These results indicate that the influence such as accelerating of color-fading by mordants and the combination with dyes is not changed by the differences between cotton and linen fibers.

Changes in Tensile Properties

The changes in tensile properties of undyed-mordanted and dyed-mordanted cotton yarns as a function of light exposure are shown in Table III. Neither G yarns nor S yarns showed any significant change in tensile properties resulting from dyeing and/or mordanting before exposure. However, the undyed- and dyed-Fe-mordanted, the undyed-Al-mordanted samples and not the dyed-Al-mordanted samples were severely damaged by light exposure. Since the breaking loads of all cotton samples retained 68~93% of the initial values even after the 500 h exposure, the state of degradation of cellulose fibers is considered to be still at the initial stage. Breaking loads of undyed-Fe-mordanted cotton fabrics were decreased to 68~75% of the initial values at the longest exposure and the changes were slightly larger than those of undyed-Al-mordanted fabrics. This was true for both G and S samples.

Table III. Changes in Tensile Properties of Undyed-Mordanted and Dyed- Mordanted Cotton Fabrics after Light Exposure.

Sample	Strength Retained (%)	Energy to Break Retained (%)	Δ Elongation at Break (%)	Sample	Strength Retained (%)	Energy to Break Retained (%)	Δ Elongation at Break (%)
G	81	74	-0.1^b	S	80	65	-5.0
G-AL1	72	76	-0.4^b	S-AL1	74	63	-0.8^b
G-AL2	73	74	-0.3^b	S-AL2	73	87	0.6^b
G-AL3	75	73	0.6^b	S-AL3	75	85^b	1.5^b
G-FE1	75	61	-3.0	S-FE1	75	74^b	3.4^b
G-FE2	69	53	-2.6^b	S-FE2	69	75	2.7^b
G-FE3	68	72	-6.0	S-FE3	68	80	0.9^b
GCUR-AL1	78	73	1.6	SCUR-AL1	81	83	5.7
GCUR-FE1	73	45	-0.4^b	SCUR-FE1	74	71	-1.4^b
GHEMA-AL1	81	50	-3.1^b	SHEMA-AL1	93	83^b	1.4^b
GHEMA-FE1	69	40	3.2	SHEMA-FE1	72	65	2.6^b

[a] Difference in tensile property between the corresponding samples unexposed and exposed for 500 h was evaluated.

[b] No significant change in tensile property of exposed sample was found in comparison to the unexposed sample at 95% confidence level.

The breaking loads of dyed-Fe-mordanted samples were reduced to 69~74% of the initial values at 500 h exposure and the strength loss of these samples was more than that of the corresponding dyed-Al-mordanted samples by 7~22%. The strength of the CUR-dyed-Al-mordanted samples showed an earlier decrease than that of the HEMA-dyed-Al-mordanted samples (Figure 1). From the fact that the color of the dyed-alum-mordanted samples faded rapidly, it is suggested that the dyes used to color the alum-mordanted samples degraded in preference to the fibers and may have favorably influenced the degradation of the fibers. In the case of CUR-dyed-Al-mordanted samples, the color completely faded so early during the exposure that this protection by the preferential degradation of the dye may have been lost after complete fading.

At the 500 h exposure, the breaking loads of dyed-Al-mordanted samples were reduced to 78~93% of the initial values, while those of dyed-Fe-mordanted samples were reduced to 69~74%. The difference in the strength loss between HEMA-dyed-Al-mordanted and HEMA-Fe-mordanted samples was significant at the 99% confidence level. A significant difference in the strength loss after 500 h exposure was not observed between CUR-dyed-Al-mordanted and CUR-dyed-Fe-mordanted samples. In the case of undyed-mordanted samples, the breaking loads of undyed-Al-mordanted samples were reduced to 72~75% after the 500 h exposure and those of undyed-Fe-mordanted samples were reduced to 68~75%. The difference in strength loss of undyed-Al-mordanted and undyed-Fe-mordanted G samples was significant at 99% confidence level. Although these differences between samples treated with Fe- and Al-mordants were not always significant (Table III), mordanting with iron salts induced a greater loss in the strength than mordanting with aluminum salts, which was commonly observed for both the undyed- and the dyed-mordanted samples. A similar influence of iron mordants was also observed in the corresponding linen samples (8). It should be noted that these affects of the mordants are observed in all the dyed-mordanted cotton and linen samples examined in spite of the difference in the dyes and types of cellulosic fibers.

Changes in the energy to break also showed the degradative effect of Fe mordants similarly to those of the breaking load (Table III). In the values of elongation at break, no clear characteristic change was observed in the data, probably owing to the fact that the fiber substrates were not so severely damaged by light exposure at this stage.

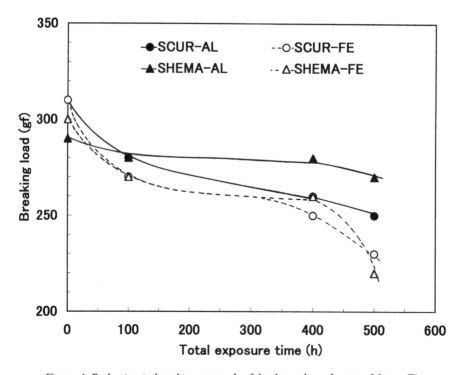

Figure 1. Reduction in breaking strength of dyed-mordanted cotton fabrics (S) after light exposure for 500h.

Conclusion

This work shows the influence of mordants and dyes on color changes and tensile properties of cotton fabrics. It also shows that the combination of mordant and dye, especially the type of mordant significantly affects color-fading and strength of fibers on light exposure. A protective effect caused by preferential degradation of dye and the degradative effect of iron salt were found in this work.

Based on this information, conservators would be advised to impose severer limits on exhibition time for Fe-mordanted fabrics. For alum- and Fe-mordanted fabrics that are already faded, there would be the further risk of strength loss by additional exposure. Light exposure is cumulative and impacts negatively upon historic cottons through visually apparent changes and alternatives in physical-mechanical properties.

Reference

1. Ranby B. In *Wood Processing and Utilization*; Kennedy, J. F.; Phillips, G. O.; Williams, P. A., Eds.; Ellis Horwood Ltd.: West Sussex, England, 1989, pp353-360.
2. Allen, N. S. In *Polymer in Conservation;* Allen, N. S.; Edge, M.; Horie, C.V., Eds.; Royal Society of Chemistry: Cambridge, England, 1992, pp 193-213.
3. Egerton, G. S. *J. Soc. Dyers and Col.* **1947**, *63*, pp 161-171.
4. Bamford, C. H.; Dewar, J. S. *Nature* **1949**, *163*, pp 214.
5. Morean, J. J.; Stonehill, H. I. *J. Chem. Soc.* **1957**, pp765-779.
6. Griffiths, J. In *Developments in Polymer Photochemistry*; Allen, N. S. Ed.; Elsevier Applied Science Publishers Ltd.: London, 1980; Vol. 106, p 145.
7. Needles, H. L.; Cassman, V.; Collins, J. In *Historic Textile and Paper Materials*; Needles, H. L.; Zeronian S. H., Eds.; Advances in Chemistry Series 212; American Chemical Society; Washington, D.C., 1986; pp 199-210.
8. Kohara, N.; Kanei, M.; Takizawa, C., Sen'I Gakkai Preprints 1998 (G), 1998, pp G-54.
9. Yatagai, M.; Magoshi, Y; Becker, M. A.; Sano, C.; Ikuno, H.; Kohara, N.; Saito, M., submitted to *Historic Textile and Paper and Polymers in Museums*.
10. Vegetable *Dyeing –Japanese Color-*; Yamazaki, Bijutsu Shuppansha: Tokyo, 1972.
11. Hon, N.-S. *J. Appl. Polym. Sc.* 1975, *19*, pp 2789-2797.

Chapter 7

Degradation and Color Fading of Silk Fabrics Dyed with Natural Dyes and Mordants

Mamiko Yatagai[1], Yoshiko Magoshi[2], Mary A. Becker[3], Chie Sano[4], Harumi Ikuno[5], Natsuko Kohara[6], and Masako Saito[7]

[1]Toyoko Gakuen Women's College, Tokyo 158–8586, Japan
[2]Department of Insect Technology, National Institute of Sericultural and Entomological Science, 1–2 Owashi, Tsukuba Ibaraki 305–8634, Japan
[3,4]Tokyo National Research Institute of Cultural Properties, Tokyo 110–8713
[5]Tokyo Gakugei University, Tokyo 184–8501, Japan
[6]Showa Women's University, Tokyo 154–8533, Japan
[7]Kyoritsu Women's University, Tokyo 101–8433, Japan

The effect of dyes and mordants on photodegradation and photofading of raw and degummed silk fabrics was investigated. Photodegradation of the fiber was assessed by measuring yarn strength loss, and photofading was expressed as the difference in color before and after light exposure. The dyed fabrics mordanted with iron showed slight fading and a large decrease in tensile strength regardless of the dye upon exposure to light. The dyed fabrics mordanted with aluminum, on the other hand, showed considerable fading, and the degree of degradation was dependent on dye. The degummed fabrics showed a greater loss in tensile strength and color change than the raw fabrics, whether treated (mordanted and/or dyed) or not, which suggests some protective effect of sericin present in raw silk against light damage to the fabrics.

Silk has been widely investigated because of its importance as a historic textile fiber. Characterization has been done on naturally or artificially aged silk, and mechanism of silk deterioration has been intensively discussed (*1-4*).

The majority of historic silk textiles in museum collections have been dyed with natural dyes of plant or animal origin with the aid of mordants, which generally have

[1]Current address: Chiba University, Chiba 263-8522, Japan.
[3]Current address: Fukui University, Fukui 910-8507, Japan.
[7]Corresponding author

been metal salts. In Japan, aluminum and iron salts have been mainly used. The lightfastness of natural dyes on fiber substrates has been the subject of many studies (5-9). It is well known that natural dyes are generally sensitive to light and their fastness is greatly affected by mordants and various environmental factors. On the other hand, only a few studies have been concerned with the degradation of fibers that usually accompanies the light-induced color change of the fabrics (10-13).

The objective of our study is to gain a better understanding of the deterioration of natural-dyed fabrics, including degradation of fibers as well as decomposition of dye molecules on the fiber. Natural-dyed fabrics of cellulosic fiber (cotton) and protein fiber (silk) have been prepared using several combinations of dye and mordant according to traditional Japanese dyeing procedures (14). The fabric samples were then exposed to a xenon lamp filtered to simulate sunlight and to accelerate the color fading and degradation in a way similar to which historic textiles were supposed to be damaged. The effect of dye, mordant and fiber substrate on the deterioration of the fabrics caused by light has been investigated by various analytical techniques. This paper presents the results for silk, and the results for cotton are described elsewhere (15).

Historically silk has been utilized with various degrees of degumming, i.e. removal of the sericin which is covering the two fibroin filaments, to obtain varieties of fabrics having different colors, lusters and fabric hands. In this work the effect of presence or absence of sericin in silk has also been studied.

Experimental

Materials

Two kinds of silk fabrics were used: a plain woven fabric made of raw silk (75.3 g/m^2, 57.3 denier warp yarn), and a degummed silk fabric (59.0 g/m^2, 43.3 denier warp yarn) obtained by degumming a portion of the raw silk fabric. Both fabrics were prepared in the National Institute of Sericultural and Entomological Science in Japan. Degumming was carried out in a degumming bath containing 3~4 g/ℓ of Marseilles soap, 0.3 g/ℓ of sodium carbonate, 1~3 g/ℓ of disodium metasilicate, 0.5 g/ℓ of sodium dithionite and 0.5 g/ℓ of a nonionic surfactant for 1 to 2 hours at the boil with a liquor ratio of 1:200. The degumming process was repeated three times and then the fabrics were rinsed and air-dried.

The metal salts used as mordants were aluminum potassium sulfate, aluminum acetate, aluminum chloride, iron (II) sulfate, iron (II) acetate, and iron (II) chloride, which were selected to observe any effects of metal cations and anions of mordant. They were all reagent grade from Wako Chemicals Co., except aluminum acetate (from Tanaka-Nao Dye Materials Co., Japan) and iron (II) acetate (from Aldrich Chemicals Co.). Two dyes found as major components in historic natural dyes were used; curcumin (C.I.75300) found in the root of turmeric, and hematoxylin (C.I.75290) found in the heart-wood of logwood. They were >95% pure and were from Merck & Co..

88

Dyeing and Mordanting Procedures

Preparation of Mordanted Fabric Samples
The mordanted fabric samples were prepared by treating silk fabrics with mordants only. Fabric pieces measuring 11×13 cm were immersed in different mordant solutions at room temperature for half an hour with a liquor ratio of 1:30. The mordant concentration was 8% on weight of fiber (owf) for aluminum mordants, and 14% owf for iron mordants. After mordanting, the fabrics were rinsed with deionized water and were allowed to air dry.

Preparation of Dyed Fabric Samples
The dyed fabric samples were prepared by repeating the dyeing and mordanting processes. Four combinations of dye and mordant were used with two dyes being curcumin and hematoxylin, and the two representative mordants being aluminum potassium sulfate (alum) and iron (Ⅱ) sulfate (copperas). First, dyeing was carried out at 80-85℃ for half an hour with 3% owf of dye and a liquor ratio of 1:30. Then the fabrics were post-mordanted under the same condition as for preparing the mordanted/undyed fabric samples. The dyeing and mordanting processes were repeated a total of three times and twice, respectively. After the procedures the dyed fabric samples were rinsed thoroughly with deionized water to remove unfixed dye, and then allowed to air dry.

Light Exposure

The fabric samples were exposed to simulated sunlight from a xenon source filtered with borosilicate inner and outer filters to simulate light exposure that actual historic objects would have been subjected to during their lifetimes. The light exposures were conducted in a Xenon Long-Life Fade-Meter FAL-25AX-HC (Suga Test Instruments Co.) according to Japanese Industrial Standard L0803 at an average black panel temperature of 63 ℃ and at an average relative humidity of 25% RH. The irradiance was 320 ± 10 W/m^2 in the range of 300-700nm. Both sides of the fabrics were exposed: first, one side was exposed for a given time period, and then the other side was exposed for the same time period. The total exposure time was 40, 100 and 400h.

Property Measurements

Color Measurements
The color of each fabric sample was measured on a Macbeth CE-7000 color spectrophotometer under Illuminant C. The color coordinates L*, a* and b* were determined for each sample, and the color differences (ΔE^*) between nonexposed and light-exposed samples were calculated using the CIE 1976 L*a*b* formula (16): $\Delta E^* = [(\Delta L^*)^2 + (\Delta a^*)^2 + (\Delta b^*)^2]^{1/2}$, where ΔL^* is the change in lightness, from lighter (+) to darker (-), Δa^* is the change in shade from red (+) to green (-), and Δb^*

is the change in shade from yellow (+) to blue (-). The unit of color difference is abbreviated CIELAB unit. The color measurements were made on both sides of the fabrics, and the values for both sides were averaged.

Tensile Testing

Breaking strength and elongation at break were measured for warp yarns removed from the fabrics at a gauge length of 20 mm and a rate of extension of 20 mm/min on a Tensilon UTM-III (Toyo Baldwin Co.). Prior to testing, the samples were conditioned at 20 ℃ and 65% RH. Fifteen yarns were tested for each sample. The standard deviation was less than 10% for the nonexposed samples and within 20% for the exposed samples.

Determination of Mordants on Fabrics

The amounts of aluminum and iron adsorbed by the fabric samples were determined by elemental analysis using inductively coupled plasma spectrometry (ICP). The fabric samples (0.3 g) were dissolved in acidic solutions and then analyzed on a Seiko SPS 1500VR. The concentrations of aluminum and iron in the sample solutions were determined using the control solutions containing 10 ppm of aluminum or iron.

Results and Discussion

Determination of Mordants on Silk Fabrics

The amounts of aluminum and iron found on the nonexposed silk fabric samples are listed in Table I. The fabrics treated with mordants, mordanted only, and the dyed fabrics (dyed three times, mordanted twice) contained 1.83-5.80 mg of aluminum or iron per gram of fabric, while negligible amounts were found on the untreated fabrics. The metals are thought to form complexes with dye and/or fibers in the treated fabrics. The treated raw samples generally contained larger amounts than the treated degummed samples. It is likely that sericin, a gum-like protein present in raw silk, absorbs more metal ions. The dyed samples contained more metals than the mordanted, undyed samples due to the repetition of the mordanting process. Furthermore, more iron was found than aluminum both in the mordanted samples and in the dyed samples. This can be ascribed in part to the higher concentration of mordant bath for iron than for aluminum.

The nonexposed silk fabrics were also subjected to the measurements of electron spin resonance (esr). A broad singlet that indicates an organic radical was observed in the esr spectra (measured at 21℃) of the dyed fabrics mordanted with aluminum, but not in the spectra of the other fabrics. This radical seems to be the same species as found in the case of cotton (*15*).

Table I. Adsorbed Aluminum and Iron by Untreated or Treated Raw and Degummed Silk Fabrics as Determined by ICP

Fabric Sample			Al (mg/g Fabric)		Fe (mg/g Fabric)	
Treatment	Dye	Mordant	Raw	Degummed	Raw	Degummed
Untreated	-	-	0.10	0.10	0.10	0.23
Mordanted/Undyed	-	KAl(SO4)2	2.00	1.83		
Mordanted/Undyed	-	FeSO4			3.03	3.53
Mordanted/Dyed	Curcumin	KAl(SO4)2	3.03	2.13		
Mordanted/Dyed	Curcumin	FeSO4			4.17	3.67
Mordanted/Dyed	Hematoxylin	KAl(SO4)2	3.43	2.47		
Mordanted/Dyed	Hematoxylin	FeSO4			5.80	4.23

Light-Induced Changes in Untreated Silk Fabrics

The changes in color and tensile properties of the untreated (unmordanted, undyed) silk fabrics caused by light exposure are shown in Table II. The untreated raw and degummed silk fabrics slightly yellowed upon light exposure, and the color difference between the nonexposed silk and the light-exposed samples increased slightly with an increase in exposure time. The degummed silk showed a slightly larger discoloration than the raw silk. In addition, photodegradation assessed by loss in tensile properties was larger in the degummed silk than in the raw silk. The loss in yarn breaking strength after 400h-exposure was ca. 38% for the raw silk, and ca. 50% for the degummed silk. The loss in elongation at break was also larger in the degummed silk. The degummed silks were more affected by light. The presence of sericin coating seems to provide some protection against light damage.

Light-Induced Changes in Mordanted/Undyed Silk Fabrics

The color and yarn breaking strength of mordanted/undyed silk fabrics and their light-induced changes after 400h-exposure are listed in Table III.

Mordanting with the aluminum mordants caused no substantial color change in the fabrics, while the fabrics mordanted with iron showed a beige color. Light exposure induced more discoloration in the aluminum-mordanted samples than in the untreated control of which discoloration after 400h-exposure is shown in Table II. The iron-mordanted samples became progressively darker and the color difference increased more markedly with exposure time. Furthermore, the degummed samples showed a larger color change than the raw samples.

Mordanting also caused slight changes in tensile properties of the samples; a decrease in yarn breaking strength and an increase in elongation at break. Both properties decreased upon exposure to light, although the loss in breaking strength was more obvious and more dependent on the fiber-mordant combination. As shown in Table II and III, all the mordanted samples showed a greater loss in strength than the untreated control. The damage was greater in the iron-mordanted samples than in the aluminum-mordanted samples. Especially, the iron-mordanted degummed samples were severely damaged. The difference of anions contained in metal salts had less effect when compared to the metals themselves. From these results, it is obvious that aluminum and iron tend to accelerate discoloration and degradation of silk.

Light-Induced Changes in Mordanted/Dyed Silk Fabrics

The color and yarn breaking strength of mordanted/dyed silk fabrics and their light-induced changes after 400h-exposure are listed in Table IV.

Turmeric, which contains curcumin as a major coloring component, is known to have poor lightfastness. As anticipated, the curcumin-dyed fabrics exhibited greater fading than the hematoxylin-dyed fabrics did for the same mordant. For each dye, the samples mordanted with aluminum showed greater fading than the samples mordanted with iron. Moreover, the degummed fabrics showed a greater degree of fading than the raw fabrics. In summary, these results indicate that the degree of

Table II. Discoloration and Changes in Yarn Tensile Properties of Untreated (Unmordanted, Undyed) Raw and Degummed Silk Fabrics Caused by Light Exposure for up to 400 hours

Total Exposure Time (h)	Color Difference (ΔE*)		Breaking Strength (gpd)		Elongation at Break (%)	
	Raw	Degummed	Raw	Degummed	Raw	Degummed
0	-	-	3.32 (0.20)	3.18 (0.22)	22.8 (2.2)	16.7 (1.7)
40	0.3	1.2	3.25 (0.14)	2.93 (0.32)	20.3 (2.8)	12.8 (2.1)
100	0.3	1.9	3.06 (0.22)	2.87 (0.19)	20.5 (3.4)	13.2 (1.8)
400	3.0	5.5	2.05 (0.19)	1.58 (0.17)	10.2 (1.6)	6.1 (1.0)

NOTE: Discoloration is expressed as the color difference (ΔE*) between nonexposed and exposed samples calculated using the CIE 1976 L*a*b* Color difference formula. Value in parentheses indicates the standard deviation.

Table III. Color and Yarn Breaking Strength of Mordanted/Undyed Raw and Degummed Silk Fabrics and Their Changes after 400 hours of Light Exposure

Fiber	Mordant	Color						Color Difference (ΔE^*)	Breaking Strength (gpd)		Strength Loss (%)
		nonexposed			400h-exposed				nonexposed	400h-exposed	
		L^*	a^*	b^*	L^*	a^*	b^*				
Raw	KAl(SO$_4$)$_2$	86.9	-0.6	8.5	85.4	-0.5	12.5	4.2	2.95 (0.17)	1.38 (0.14)	53
	Al(CH$_3$COO)$_3$	86.5	0.1	8.6	84.3	0.3	12.7	4.7	2.89 (0.23)	1.01 (0.13)	65
	AlCl$_3$	87.1	-0.4	8.8	84.6	0.2	12.9	4.8	3.00 (0.25)	1.13 (0.13)	62
	FeSO$_4$	66.2	5.2	23.0	52.9	5.1	19.3	14.3	3.04 (0.17)	0.51 (0.08)	83
	Fe(CH$_3$COO)$_2$	67.7	6.0	22.3	55.4	5.7	19.3	12.9	3.06 (0.12)	0.82 (0.07)	73
	FeCl$_2$	66.6	5.2	21.4	53.6	5.7	19.7	13.4	2.96 (0.10)	0.62 (0.09)	79
Degummed	KAl(SO$_4$)$_2$	93.6	-0.5	3.2	90.2	-0.8	8.8	6.6	2.83 (0.21)	1.52 (0.14)	46
	Al(CH$_3$COO)$_3$	93.7	-0.5	3.2	91.7	-1.0	9.0	6.2	3.16 (0.23)	1.37 (0.15)	57
	AlCl$_3$	93.6	-0.5	3.2	91.4	-1.1	9.6	6.6	2.83 (0.20)	0.87 (0.12)	69
	FeSO$_4$	76.4	2.6	15.9	60.6	3.0	15.9	16.0	2.87 (0.19)	0.06	98
	Fe(CH$_3$COO)$_2$	77.8	2.8	15.6	60.7	2.7	14.8	17.4	2.94 (0.22)	0.06	98
	FeCl$_2$	75.5	3.5	18.2	61.9	3.5	16.1	14.0	2.91 (0.21)	0.11	96

NOTE: $\Delta E^* = [(\Delta L^*)^2 + (\Delta a^*)^2 + (\Delta b^*)^2]^{1/2}$, where ΔL^*, Δa^* and Δb^* are the differences of the color coordinates, L^*, a^* and b^* values between nonexposed and 400h-exposed samples. Value in parentheses indicates the standard deviation.

94

Table IV. Color and Yarn Breaking Strength of Mordanted/Dyed Raw and Degummed Silk Fabrics and Their Changes After 400hours of Light Exposure

Fiber	Dye	Mordant	Color						Color Difference (ΔE^*)	Breaking Strength (gpd)		Strength Loss (%)
			nonexposed			400h-exposed				nonexposed	400h-exposed	
			L^*	a^*	b^*	L^*	a^*	b^*				
Raw	Curcumin	KAl(SO$_4$)$_2$	63.2	18.2	71.9	68.6	8.1	28.6	45.4	2.99 (0.15)	1.85 (0.12)	38
	Curcumin	FeSO$_4$	28.8	10.4	13.4	34.8	9.6	20.5	7.8	2.90 (0.14)	1.65 (0.10)	43
	Hematoxylin	KAl(SO$_4$)$_2$	19.2	7.3	-6.6	22.8	5.9	-2.0	7.7	2.85 (0.12)	2.60 (0.17)	9
	Hematoxylin	FeSO$_4$	14.6	1.4	1.3	17.6	0.4	-0.2	2.3	2.81 (0.12)	1.75 (0.14)	38
Degummed	Curcumin	KAl(SO$_4$)$_2$	70.5	17.0	82.1	81.5	4.5	17.7	67.1	2.76 (0.20)	0.31 (0.06)	89
	Curcumin	FeSO$_4$	35.3	15.1	22.3	44.9	7.6	24.1	13.6	2.72 (0.15)	0.36 (0.10)	87
	Hematoxylin	KAl(SO$_4$)$_2$	26.1	11.7	-19.1	45.4	7.0	0.7	28.5	2.86 (0.17)	1.16 (0.19)	59
	Hematoxylin	FeSO$_4$	18.0	1.5	-2.9	32.3	0.8	5.1	13.3	2.56 (0.22)	0.12 (0.04)	95

NOTE: $\Delta E^*=[(\Delta L^*)^2+(\Delta a^*)^2+(\Delta b^*)^2]^{1/2}$, where ΔL^*, Δa^* and Δb^* are the differences of the color coordinates, L^*, a^* and b^* values betwee nonexposed and 400h-exposed samples. Value in parentheses indicates the standard deviation.

color fading of the dyed silk fabrics is dependent not only on dye itself but also on mordant, and the nature of the fiber substrate (in this case, presence or absence of sericin).

The dyeing process caused a decrease in the breaking strength of the samples as did the mordanting process. Light exposure induced the same or greater degree of strength loss in all the samples compared to the untreated control except raw silk dyed with hematoxylin and mordanted with aluminum. The degradation was much more noticeable in the degummed samples than in the raw samples. In both fiber substrates, the degradation in the samples mordanted with iron was severe irrespective of the dye. On the other hand, the degradation in the samples mordanted with aluminum was more dependent on the dye. The hematoxylin-dyed samples showed less degradation than the curcumin-dyed samples.

It is hypothesized that if decomposition of the dye is preferential to degradation of the fiber substrate, the loss in physical properties is delayed. In other words, the fiber substrate is protected from degradation while dye fading is in progress. This seems to be the case for the combination of hematoxylin and aluminum, in which a high degree of dye fading and smallest decrease in strength were observed. On the other hand, the least lightfast samples, dyed with the combination of curcumin and aluminum, showed a high degree of degradation also. In this combination, the process of dye fading was very rapid and terminated in a very short period. After the 400h-exposure, the yellow color of the curcumin-dyed/aluminum-mordanted fabrics had been almost lost. It is likely that dye fading ended very soon and the fiber substrate deterioration proceeded immediately.

The light-induced changes in color and breaking strength in the dyed silk fabrics are compared with the results of the cotton fabrics (15) in Table V. Both sets of the fabrics were dyed and mordanted, and then subjected to light exposure under the same conditions for up to 400 hours. With an increase in exposure time, the silk fabrics and the cotton fabrics faded to the same extent. Namely, the color change of the fabrics was dependent primarily on the dye-mordant combination. The loss in breaking strength was much greater in silk than in cotton, although a smaller loss in strength in the hematoxylin-dyed/aluminum-mordanted fabric was commonly observed in both fiber substrates.

Conclusions

The degree of photodegradation and photofading of the natural-dyed silk fabrics was dependent not only on the dye-mordant combination but also on the presence or absence of sericin on the silk fibers. Regardless of the dye applied, the fabrics mordanted with iron showed minimal fading, while exhibiting an increasing loss in physical properties as a function of aging. However, for the aluminum-mordanted samples the degree of fading was dependent on the dye. It was apparent from the aluminum-mordanted dyed samples that for certain dye-mordant combinations photofading preceded the fiber substrate deterioration. Only after the fading had reached completion was a loss in the physical properties observed. In addition, the results indicate some protection to the fibers from light damage due to the presence of a sericin coating.

Table V. Comparison of Changes in Color and Yarn Breaking Strength of Silk Fabrics (Degummed) and of Cotton Fabrics (Scoured and Bleached) Dyed and Mordanted, and Exposed to Light for 400 hours Under the Same Conditions.

Dye	Mordant	Color Difference (ΔE^*) between Nonexposed and 400h-exposed Samples		Retention of Breaking Strength after 400h-exposure (%)	
		Silk	Cotton	Silk	Cotton
Curcumin	KAl(SO$_4$)$_2$	67.1	64.9	11	81
Curcumin	FeSO$_4$	13.6	18.6	13	81
Hematoxylin	KAl(SO$_4$)$_2$	28.5	26.7	41	97
Hematoxylin	FeSO$_4$	13.3	5.3	5	87

NOTE: $\Delta E^* = [(\Delta L^*)^2 + (\Delta a^*)^2 + (\Delta b^*)^2]^{1/2}$, where ΔL^*, Δa^* and Δb^* are the differences of the color coordinates, L^*, a^* and b^* values between nonexposed and 400h-exposed samples. Details of the results for cotton are given elsewhere (15).

Photodegradation and dye fading of the dyed silk fabrics and dyed cotton fabrics progressed in a similar manner, although silk was more altered by light exposure. This information on dye-mordant combinations should be useful to those who are responsible for the care and preservation of natural-dyed fabrics. Even those samples, which have not been subjected to the extreme "weighting" processes commonly found in Occidental fabrics, can be at substantial risk to physical damage due to light exposure. If a fabric is known to be mordanted with iron, its color may show little variation while there could be a substantial loss in its physical properties for a given exposure. Photofading of aluminum-mordanted fabrics, particularly those dyed with curcumin, appears to precede actual loss in physical properties. So if an object is already faded, it will be a risk to further damage if subjected to additional exposure. Furthermore, it should be kept in mind that exposures are cumulative when considering how an object will be treated, stored or exhibited.

References

1. Hirabayashi, K. *Scientific Papers on Japanese Antiques and Art Craft* **1981**, *26*, 24-34.
2. Hirabayashi, K.; Takei, M.; Kondo, M. *Archaeology and Natural Science* **1984**, *17*, 61-72.
3. Hersh, S.P.; Tucker, P.A.; Becker, M.A. In *Archaeological Chemistry IV*; Allen, R.O., Ed., American Chemical Society: Washington, DC, 1989; pp.429-449.
4. Becker, M.A.; Magoshi, Y.; Sakai, T.; Tuross, N.C. *Stud. Conserv.* **1997**, *42*, 27-37.
5. Padfield, T.; Landi, S. *Stud. Conserv.* **1966**, *11*, 181-196.
6. Duff, D.G.; Sinclair, R.S.; Stirling, D. *Stud. Conserv.* **1977**, *22*, 161-169.
7. Kashiwagi, M.; Yamazaki, S. *Scientific Papers on Japanese Antiques and Art Crafts* **1982**, *27*, 54-65.
8. Bowman, J.G.; Reagan, B.M. *Stud. Conserv.* **1983**, *28*, 36-44.
9. Crews, P.C. *Stud. Conserv.* **1987**, *32*, 65-72.
10. Watase, H.; Oguchi, T.; Horikawa, S.; Sasaki, E. *J.Japan Research Association for Textile End-Uses* **1978**, *19*, 69-72
11. Needles, H.L.; Cassman, V.; Collins, M.J. In *Historic Textile and Paper Materials*; Needles, H.L.; Zeronian, S.H., Eds.; Advances in Chemistry Series No.212; American Chemical Society: Washington, DC, 1986; pp199-210.
12. Urabe, S.; Yanagisawa, M. *Scientific Papers on Japanese Antiques and Art Crafts* **1988**, *33*, 10-17.
13. Urabe, S.; Yanagisawa, M. *Scientific Papers on Japanese Antiques and Art Crafts* **1989**, *34*, 11-19.
14. Yamazaki, S. *Vegetable Dyeing - Japanese Color*; Bijutsu Suppansha: Tokyo, 1972.
15. Kohara, N.; Sano, C.; Ikuno, H.; Magoshi, Y.; Becker, M.A.; Yatagai, M.; Saito, M. submitted to *Historic Textile and Paper and Polymers in Museums.*
16. *AATCC Technical Manual;* American Association of Textile Chemists and Colorists: Research Triangle Park, NC, 1995, vol.70, pp 274-275.

Chapter 8

Measuring Silk Deterioration by High-Performance Size-Exclusion Chromatography, Viscometry, and Electrophoresis

Season Tse and Anne-Laurence Dupont[1]

Canadian Conservation Institute, 1030 Innes Road, Ottawa,
Ontario K1A 0M5, Canada

Deterioration of silk textiles in museums due to improper
storage, display conditions and treatments is a concern for
many conservators and collectors. To better preserve these
textiles, early detection and quantitative assessment of
damage is essential. This paper evaluates the sensitivity of
three analytical techniques developed for measuring silk
deterioration: high performance size exclusion
chromatography, viscosity measurements, and sodium
dodecylsulphate polyacrylamide gel electrophoresis. Silk
habutae were artificially light aged and treated by immersion
in deionized water, ethanolic solution of sodium
borohydride, aqueous solutions of sodium dodecylsulphate,
and protease. The results showed that all three techniques are
very sensitive to the small changes in silk molecular weight
resulting from deterioration, and that they complement each
other by providing different types of information.

For many years, research into the care of degraded silk textiles have focussed
on mechanisms of *Bombyx mori* silk deterioration (1,2), effects of storage and display
environments (3,4), methods of consolidation (5,6), effects of weighting (7,8,9), and
cleaning (10,11,12). More recently the long-term effects of various wet-cleaning

[1]Current address: National Gallery of Art, DCL-Scientific Research Department, Washington,
DC 20565.

reagents, treatment techniques, exposure to UV-filtered high intensity light (400-800nm), and UV-attenuated fluorescent lights have also become a concern. To evaluate the impact of these processes effectively, it is necessary to establish analytical methods that are both sensitive and quantitative to small changes in silk as a result of deterioration. Traditionally, damage in silk textiles is measured by fading of dyes or yellowing of the fabric, loss of surface properties such as gloss or luster, and loss of mechanical strength. Many of these techniques are suitable for assessment of end-use properties, but they are not sufficiently sensitive to allow early detection of damage. This study focussed on techniques that are sensitive to polymer molecular weights changes because there is already a well-established relationship between molecular weight and certain physical manifestations of deterioration such as tensile strength, tenacity, toughness, chemical resistance, and abrasion resistance (13). Three techniques were investigated: viscometry, high performance size exclusion chromatography (HPSEC), and sodium dodecylsulphate polyacrylamide gel electrophoresis (SDS-PAGE).

Viscometry is one of the most used techniques for studying molecular weight of polymers because of its simplicity, sensitivity and reproducibility. At constant temperature, the viscosity of a dilute macromolecular solution is a function of its molecular size and shape, and is proportional to its molecular weight (14,32). In an earlier study, Tweedie showed that the technique is quantitative for assessing silk fibroin deterioration, and could detect initial photodegradation in silk, resulting in random chain scission, before any major loss of strength (15).

Next to viscometry, size exclusion chromatography (SEC) is the most common technique for studying molecular weight of polymers (13). In SEC, polymer molecules in the mobile phase are separated by their size in the column's gel matrix. The use of HPSEC to characterize the size and molecular weight of proteins is relatively recent (16,17,18,19). Conventional SEC has been used to study molecular weight of silk fibroin extracted from the silk gland (20,21) and deteriorated silk (2). The use of HPSEC for measuring silk degradation has been reported by Howell as promising for routine analysis (22).

Electrophoresis, using SDS polyacrylamide gels (SDS-PAGE), is commonly used for separating proteins according to their molecular weights. Proteins are denatured by heating at 100°C in the presence of a reducing agent, ß-mercaptoethanol, and a surfactant, sodium dodecylsulphate, SDS. The reducing agent removes interchain disulphide linkages, freeing the protein sub-units. The SDS then binds to the polypeptides forming rod-shaped protein-SDS complexes with a consistent diameter and a strong negative charge. The protein-complexes are separated in the acrylamide gel matrix according to its molecular weight when an electric field is applied (23). The molecular weight of the protein is determined by comparison with the relative mobility (R_f) of protein molecular weight markers applied on the same gel, using a calibration curve. SDS-PAGE has been used to study the molecular weight of the silk subunits (24,25,26), and it is included in this study to gain additional information about the changes in the subunits as silk degrades.

All three techniques used in this study measure relative rather than absolute molecular weights, but their sensitivity makes them useful for monitoring changes in molecular weights. Both viscometry and HPSEC are sensitive to changes in hydrodynamic volume (V_H), which is a function of molecular size and shape of polymers in solution. An important assumption is that a major decrease in intrinsic viscosities or increase in SEC elution times is interpreted as a decrease in molecular weight caused by increased chain scission, and that the conformation of the silk protein in solution does not change to any great extent. This paper compares the sensitivities between HPSEC and viscometry, and evaluates the usefulness of SDS-PAGE for measuring silk deterioration.

Experimental

Commercially degummed silk habutae (Testfabrics Inc; #609; 36 g/m^2) was used for all experiments without further processing. Silk fabrics were exposed to different dosages of xenon-arc light to create a set of systematically aged samples. The light-aged silk was subjected to four wet-cleaning treatments: water, aqueous solutions of sodium dodecylsulphate, protease, and ethanolic solution of sodium borohydride. Silk samples were analyzed after dissolution in lithium thiocyanate solvent.

Artificial Aging in Weather-Ometer

An Atlas Weather-Ometer (65-WRC 123) equipped with xenon-arc lamp was used. Borosilicate inner and sodalime outer filters were used to simulate sunlight through window glass exposure. Black panel temperature was maintained at 63±3°C and relative humidity (RH) at 30 ±5% according to AATCC Test Method 16-1990(27). The lamp was calibrated and maintained at 1 W/m^2 monitored at 420 nm. Five pieces of silk habutae (approximately 75 x 40 cm) were mounted directly on the rotating rack using large paper clips. The silk was isolated from the metal clips by strips of acid-free mat board. The samples were irradiated at 100, 200, 300, 500 or 750 kJ/m^2 monitored at 420 nm, corresponding to exposure times ranging from 28 to 209 h(2). Part of the 750 kJ/m^2 sample was sandwiched between a 400 nm cut-off UV filter during exposure. A separate piece of silk habutae (40 x 5 cm) was exposed at 1000 kJ/m^2. One third of the piece was masked with acid-free matboard, one third was covered by UV-filter, and one third was exposed .

Treatments of Light-Aged Silk

Water used for all treatment and solution preparation was purified by reverse osmosis using a Millipore RO-10 unit and polished by Milli-Q Plus cartridges. The specific resistance of water was 18 MΩ. Each of the five Weather-Ometer aged pieces along with an unaged control were cut into five sections (15 x 40 cm; ~1.7 g each). One of the five sections was used as an untreated control. One piece from each of the unaged and light-aged silk (total of ~10 g) was immersed at the same time in one of the four solutions: water, sodium borohydride (NaBH$_4$) 0.5 M(w/v) in ethanol, sodium dodecylsulphate (SDS) 0.2% (w/v) in water, or protease (Pronase E; Sigma Type XIV; EC 3.4.24.31) 1 mg/mL in 0.05 MTris buffer (TRIZMA® pH 8.3 preset crystals; Sigma). A ratio of 1 g silk:100 mL solution was used, therefore approximately 1 L of each of the treatment solutions was required. The silk fabrics were immersed for 1 h in each of the treatment solutions, followed by four water rinses of 8 to 10 min each. Fabrics were air-dried for 1 week between blotters prior to analyses.

Preparation of Lithium Thiocyanate Solution

All molecular weight determinations require the polymer to be in solution, but it is crucial to choose a solvent that does not degrade silk during dissolution. Lithium thiocyanate (LiSCN) has long been recognized as the least degradative solvent to silk (28,29,30,31) compared with other solvents such as lithium bromide (Swiss Standard (32), lithium iodide (33) and zinc chloride (15,33). However, its use as a solvent for viscometry is not common in recent silk studies. To confirm the stability of silk in LiSCN and also to determine the optimum dissolving time for silk, preliminary measurements were carried out where the reduced viscosities of unaged and aged silk, in 6.67 M LiSCN at room temperature, were monitored from 1 to 28 h. The viscosities, measured at 30°C, were found to decrease for the first 8 h and remain unchanged up to 28 h. This initial decrease in viscosity was assumed to result from dissolution and disentanglement of the silk protein chains in solution, because degradation would cause the viscosities to continue to decrease with time. The optimal time for dissolution, 16 h, was chosen based on these results.

Lithium thiocyanate (LiSCN) solution (~10 M) was prepared by dissolving the content of a 100 g bottle of LiSCN·xH$_2$O (65.02 g/mol) in 71 mL of water. The final volume was 145 mL. The solution was allowed to stand overnight and centrifuged for 10 min at 3000 rpm. The clear supernatant, referred to as 10 M LiSCN, was stored in a brown polypropylene bottle and was used for dissolving silk. Since LiSCN is very hygroscopic, the amount of water associated with each mole of dry LiSCN may vary. To eliminate variations in properties of silk solutions, a single batch (145 mL) of 10 Msolution was prepared for all the analyses.

Viscometry

The viscosity of the solvent and silk solutions were measured using a Cannon-Manning semi-micro viscometer for transparent liquids (#100; Anachemia Science). The solvent, 6.67 M LiSCN, was prepared by dilution from 10 M LiSCN; 0.500 mL of solvent or silk solutions was used for measurement. The viscometer and the solution were immersed in a water bath maintained at 30.0±0.1°C, and the efflux time, t(s) was measured to an accuracy of ±0.1 s.

Silk fabrics were cut into 1.5 cm^2 and weighed (±0.1 mg) into 10 mL Erlenmeyers flasks; 4.00 mL of 10 M LiSCN was added to the weighed silk. The solutions were allowed to stand for 2 to 5 min, followed by 1 hour stirring or until the solution became clear. After leaving the silk solution unstirred at room temperature for approximately

16 h, 2.00 mL of water was added to achieve a 6.67 M LiSCN. This solution was centrifuged for 10 min at 3000 rpm and was used as a stock solution for viscosity measurements or further dilutions.

Intrinsic viscosity ($[\eta]$; dL/g) of each of the unaged and aged silks was first determined by the extrapolation method. In the extrapolation method, the stock solution ($C_{100\%}$) and two other dilutions were required. The first dilution ($C_{66.7\%}$) was prepared by adding 1.00 mL of 6.67M LiSCN to 2.00 mL of stock ($C_{100\%}$) and the second dilution ($C_{33.3\%}$) by adding 2.00 mL of 6.67M LiSCN to 1.00 mL of stock ($C_{100\%}$). Intrinsic viscosity of a silk solution was the y-intercept of a linear plot between the reduced viscosities (η_{spc}) of the three solutions vs. the corresponding concentration (c; g/dL). The calculation of the solution concentration, c, was based on the dry weight of the silk assuming a 10% moisture content (33). Reduced viscosity (η_{spc}; dL/g) is defined as: $\eta_{spc}= \eta_{sp}/c$, where η_{sp} is the specific viscosity and is obtained from the efflux time of the solvent (t_0) and of the solution (t): $\eta_{sp} = (t-t_0)/t_0$ (34).

Intrinsic viscosity of replicate silk solutions were also determined from the Schulz-Blaschke equation (35): $[\eta] =\eta_{spc}/(1+k \times \eta_{sp})$. The coefficient k for each silk sample was calculated using the extrapolated intrinsic viscosity. The intrinsic viscosities of replicate solutions were obtained from η_{spc} and η_{sp} calculated from the measured efflux times of a single concentration silk solution, usually that of the $C_{33.3\%}$. This was done to reduce the amount of samples required and time of analysis.

HPSEC

A Waters Chromatography System was used, with two Zorbax® columns GF-250 and GF-450(36) connected in series and a combined molecular mass separation range between 4.0 to 900 kilo Daltons (kDa). The column temperature was maintained at 30°C, with a flow rate of 1 mL/min and column pressure remained at 1000 psi. Elution profiles were monitored at 280 nm using a photo diode array detector. Millenium software was used for data handling and calculation of molecular weight averages based on the elution times of protein molecular weight standards. A pH 7, 0.5 M tris-urea buffer, was used as the mobile phase(37). Prior to separation, the buffer was vacuum degassed using 0.2 µm PTFE Whatman membrane filters.

Silk pieces (~8.0 mg; 1x1 cm) were dissolved in 0.50 mL of 10 M LiSCN and allowed to stand overnight. Water (0.50 ml) was added to the dissolved silk. This solution was centrifuged and filtered using a 0.45 µm PTFE syringe filters and 50 µL of the sample solution was injected. The time required for each separation was 40 min with 15 min of equilibration between sample injections. Protein molecular weight standards from Sigma-Aldrich (669 to 6.5 kDa) were dissolved in the mobile phase (1.0 mg/mL), and 50 µL of each was used to calibrate the columns and verify column performance. Blue dextran (2,000 kDa) was used to determine the void volume.

SDS-PAGE

A Multiphor II horizontal electrophoresis system, precast SDS Excel Gel

Homogeneous 15 gels, with a separation range between 4 and 170 kDa, and ExcelGel SDS buffer strips from Pharmacia were used for separation. Molecular weight markers (Low Molecular Weight Standard mixture, 94 to 14.4 kDa; 50 μg; from Pharmacia) was used for calibration.

Silk pieces (~50.0 mg; 1x1 cm) were dissolved in LiSCN (0.50 mL; 10 M). Water (100 μL) was added to the dissolved silk and the solution was dialyzed against water overnight using benzoylated cellulose dialysis tubing, with a molecular weight cut-off of 1.2 kDa (Sigma-Aldrich). Seprasol II (100 μL; Iss-Enprotech), a denaturation solution containing glycerol, SDS, tris buffer and bromphenol blue tracking dye was added to the dialyzed silk (50 μL). The solution was denatured at 100°C for 30 min. After denaturation the silk solution became a gel. Urea, to a final concentration of 9 M, was added to solubilized this gel. Protein standards were prepared by adding water (1.00 mL) directly to the vial. Concentration of each protein in the mixture was 0.1 μg/μL. Protein standards were denatured in the same manner as the silk solutions. After denaturation 20 μL protein standards and 20 μL silk solutions (~ 350 μg silk) were pipetted onto the acrylamide gel for separation. The temperature of the cooling plate was maintained at 10°C using thermostatic circulator (MultitempII 2219; LKB); maximum voltage was set at 600 V, and current set at 20-30 mA. This varies for different experiments. The number of volt-hours (VH) ranges from 900 to 950, with approximate running time between 2.5 and 3 h.

The gel was fixed for 1-2 h in a solution of methanol (150 mL), trichloroacetic acid (57.0 g), sulphosalicylic acid (17.0 g) and water (350 mL); stained for 2 hours in a staining solution made up of Coomassie Brilliant Blue R250 (1.25 g) or G250, methanol (230 mL), glacial acetic acid (40 mL) and water (230 mL). The gel was destained in a solution of methanol (250 mL) and of glacial acetic acid (80 ml) made up to 1000 mL with water, until the background became clear (approximately 24 hours). After destaining, the gel was soaked for 30 min to 1 h in a preserving solution of glycerol (99.5%; 25 mL) made up to 250 mL with destaining solution. Finally, the gel was covered with a cellophane preserving sheet (Pharmacia) and allowed to dry at room temperature.

Results and Discussions

Instrinsic Viscosities of Weather-Ometer Aged and Treated Silk

Intrinsic viscosities of light-aged and treated silk are summarized in Table I. Compared with the unaged control, there was a drastic decrease in [η] with the 100 kJ/m^2 (28 h) exposed sample, suggesting a major decrease in the polymer size in solution. The amount of change decreases with further aging. The large initial decrease suggests that the technique is capable of detecting damage much earlier than 28 hours of exposure, but it also becomes much less sensitive to change as silk becomes more degraded. The protection of 400 nm cut-off UV filters is evident in the 750 kJ/m^2 exposed samples. In comparison to the sample without the UV filter, the [η] of the protected sample is much

Table 1 Intrinsic Viscosities of Silk in Lithium Thiocyanate Solution at 30°C and HPSEC Averaged Molecular Weights: Before and After Weather-Ometer Exposure and Treatment

Treatment			*Irradiance Energy (kJ/m²) at 420 nm*				
	Unaged	*100*	*200*	*300*	*500*	*750*	*1000*
Untreated control							
[η]; dL/g	0.404±0.010	0.242±0.011	0.198±0.008	0.165±0.006	0.128±0.003	0.115± 0.005	-
k	0.42	0.59	0.71	0.67	0.81	0.81	-
M_n(kDa)	69.2±5.8	40.4±2.9	34.3±1.7	29.6±2.0	23.4±1.1	19.9±1.2	21.2
M_w(kDa)	346±29	176±19	137±15	113±14	74.8±6.2	54.2±7.1	65.8
M_p(kDa)	496±51	121±20	86.5±13.3	49.0±8.2	29.1±2.8	19.7±3.5	20.7
Untreated control + 400 nm cut off UV-Filter							
[η]; dL/g	-	-	-	-	-	0.248±0.034	-
k	-	-	-	-	-	0.45	-
M_n(kDa)	-	-	-	-	-	-	53.9
M_w(kDa)	-	-	-	-	-	-	255
M_p(kDa)	-	-	-	-	-	-	270
Deionized water washing (Water)							
[η]; dL/g	0.402±0.007	0.225	0.167	0.180	0.114	0.108	-
k	0.38	0.59	0.67	0.66	0.86	0.86	-
Sodium borohydride (BH₄) in ethanol; 0.5M							
[η]; dL/g	0.461	0.247	0.203	0.177	0.121	0.115	-
k	0.36	0.51	0.57	0.49	0.87	0.87	-
M_n(kDa)	68.8	44.8	34.9	-	-	25.8±3.0	-
M_w(kDa)	378	204	140	-	-	80.1±1.4	-
M_p(kDa)	560	158	79.7	-	-	31.4±5.4	-
0.2% w/v SDS (SDS)							
[η]; dL/g	0.367	0.220	0.163	0.158	0.126	0.114	-
k	0.56	0.60	0.74	0.68	0.86	0.87	-
M_n(kDa)	65.7	41.7	29.7	-	-	23.3±1.2	-
M_w(kDa)	340	182	107	-	-	70.7±0.3	-
M_p(kDa)	496	133	52.1	-	-	27.1±2.6	-
Protease enzyme (Enz)							
[η]; dL/g	0.479±0.012	0.250 ±0.001	0.209±0.002	0.157± 0.005	0.139±0.001	0.118±0.003	-
k	0.29	0.44	0.45	0.57	0.68	0.85	-
M_n(kDa)	66.9	43.4	34.2	-	-	26.7±3.2	-
M_w(kDa)	371	198	138	-	-	86.4±1.4	-
M_p(kDa)	550	147	76.6	-	-	34.3±7.3	-

NOTE1: k value were calculated from the Schulz-Blaschke equation using extrapolated intrinsic viscosity. M_n is the calculated number average molecular weight, M_w is the weight average and M_p is the peak molecular weight.

NOTE2: Data with standard deviations are based on an average of 3-8 replicates. Values without standard deviation are obtained from single measurements.

higher, similar to the [η] of 100 kJ/m² sample. However, there is still a decrease in viscosity compared with the unaged control. This can be attributed to damage caused by a combination of heat generated during aging or UV-filtered high intensity light. The results of wet-cleaned silk showed that there were subtle but consistent differences in [η] resulting from each treatment and these differences are most obvious with unaged silk and becomes less measurable with aged silks. For example, water immersion caused negligible change in [η]. Both borohydride and protease treatments resulted in slightly higher viscosities, and SDS treatments lead to consistently lower [η]. While these results are preliminary, the repeatability of the technique indicates that the data are reliable.

HPSEC of Weather-Ometer Aged and Treated Silk

Since HPSEC is sensitive to both molecular weight and shape of molecules in solution, accurate molecular weight determination requires that the molecules to be analyzed have the same conformation as the standards used for calibration (17). Since globular protein standards were used, they would have a different conformation from silk protein in solution, which is assumed to have random coil conformation (38). The calibration curve used for molecular weight calculations has a correlation coefficient (R^2) of 0.98 and the molecular weight averages calculated using this calibration curve, can only be considered apparent molecular weight averages. These calculated values are also expected to be larger than the actual molecular weight of silk because random coil is less compact than the globular proteins. It is with this understanding that numerical values of molecular weight averages are used to quantify changes in the silk before and after aging and treatments, in addition to qualitative comparisons of chromatograms.

Using HPSEC, the stability of silk in ~5 M LiSCN solvent was monitored immediately after dissolution up to 2 weeks. The greatest shift of chromatograms from low to high retention time occured within the first 24 h, and changed very slowly in the remaining 2 weeks. The more degraded silk showed less change with time. The chromatograms of unaged and Weather-Ometer aged silk are shown in Figure 1. The calculated weight average (M_w), number average (M_n) and peak molecular weight (M_p) values are summarized in Table I. The solvent lithium thiocyanate elutes at approximately 27.5 min. In the calculation of the molecular weight averages, exponential skim baseline correction was applied to correct for the contribution of the solvent. The weight-average molecular weight (M_w) is calculated as:
$M_w = \Sigma_{i=1}^{N}(h_i \, M_i) / \Sigma_{i=1}^{N} h_i$, where h_i is the SEC curve height of the ith fraction, and the M_i is the molecular weight of the ith fraction; the number-average molecular weight (M_n) is calculated as: $M_n = \Sigma_{i=1}^{N} h_i / \Sigma_{i=1}^{N}(h_i / M_i)$,. The peak molecular weight M_p is the molecular weight of the fraction at the peak of the SEC curve. As high molecular weight species have greater contribution to M_w, and low molecular weight species have greater influence on M_n, the calculated values of M_w are always larger than M_n, except for monodisperse systems where the two values are identical. The ratio M_w/M_n is a measure of the polydispersity of the polymer (33). Of the three calculated values, M_p was found

to be the most sensitive to changes from silk degradation and least affected by contribution from the solvent. The molecular weight of the bulk of the silk fibroin have been reported to be between 350 and 450 kDa based on sedimentation analysis, SEC and electrophoresis of native silk (20,24,39,40,41). Figure 1 shows that unaged silk habutae had an average M_p = 496 kDa and M_w=346 kDa, which are in agreement with literature values. As the silk ages the chromatogram shifted to a longer retention time, and the 750 kJ/m^2 samples had an average M_p of 19.9 kDa. With increase photodegradation, the silk chromatograms exhibit a significant right-hand skew of the peak. All lightaged silks showed broader distribution of high molecular weight fractions as compared with unaged silk. This shows that as the fibroin undergo random chain cleavage, the high molecular weight fractions are affected more than the low molecular weight portion. The changes in HPSEC molecular weight averages are in excellent agreement with viscosity results. There is an initial drastic decrease in M_p of the 100 kJ/m^2 sample compared with the unaged control, and gradually less decrease with further aging. The correlation coefficient (R^2) between [η], and both M_w and M_n is 0.998. There is a curvilinear relationship between M_p and [η], showing that for more degraded silks, the proportion of change between [η] and M_p are similar, and that there are bigger changes with M_p in the initial stages of degradation, therefore M_p is more sensitive at this stage. Even though these molecular weight averages are not absolute values, the results show an excellent empirical relationship between [η] and the HPSEC molecular weight averages for light-aged silk.

Figure 2 shows the chromatograms of 1000 kJ/m^2 silk with and without the protection of UV filter. Compared with the masked control (M_p=347 kDa), the sample protected by the UV filter showed only a slight shift to a lower molecular weight (M_p=273 kDa), whereas the unprotected sample had a substantially lower molecular weight (M_p=20.7 kDa). The chromatograms of treated samples show no major shifts compared with the untreated controls. Minor shifts were observed with the unaged - treated silk.

SDS-PAGE of Weather-Ometer Aged and Treated Silk

Silk protein is believed to consist of a heavy chain (H-chain) and a light chain (L-chain) associated via disulphide bonds (20,39). The H-chain, which accounts for 93% of the silk fibroin, is composed mainly of crystalline portions with small amino acids (-Gly-Ser-Gly-Ala-Gly-Ala)$_n$, interrupted by amorphous regions composed of bulkier residues. The L-chain accounts for the remaining 7% of the fibroin (25,42), and is composed mainly of amino acids with acidic and bulky side-chains. The molecular weight of the H-chain has been estimated to be between 325 and 450 kDa (41,42), the L-chain to be between 25 and 30 kDa (25,26,42). Under light aging conditions similar to our experiments, preferential deterioration of the L-chain and amorphous regions of the H-chain is believed to occur accompanied by loss of corresponding amino acids (43,44).

Figure 3 shows the electrophoretic patterns of light aged silk. The bulk of the silk

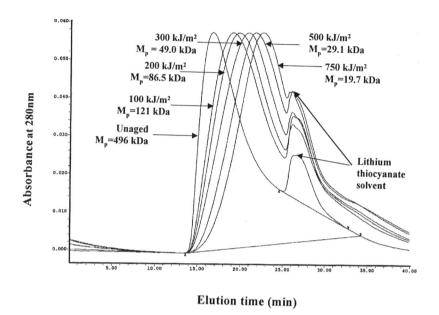

Figure 1 HPSE Chromatograms of Silk Light Aged in Weather-Ometer

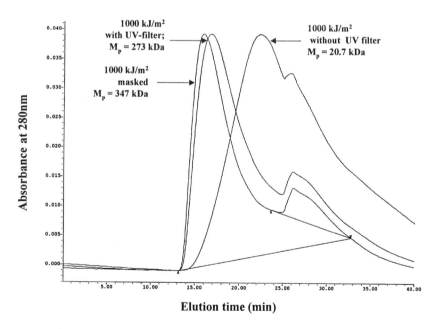

Figure 2 HPSE Chromatograms of Silk Light Aged with and without UV Filter

protein, with approximate molecular weight of 400 kDa, is too large to enter the gel to any great extent. The main band at approximately 28-30 kDa (refer to as '28kDa' in subsequent paragraphs), seen in the unaged silk, is assigned to the L-chain. With light aging, this band decrease in intensity and is lost at light exposures higher than 300 kJ/m². The protection of UV-filter is shown in Figure 4. Compared to the unaged silk, the 28 kDa band intensity of the 750 kJ/m² sample with UV-filter is only slightly diminished, whereas the unprotected 750 kJ/m² sample showed a total loss of the 28 kDa band.

The electrophoretic patterns of treated silk are also shown in Figure 4. The water, borohydride, and SDS treated silk showed very little difference in the 28 kDa band intensity as a result of treatment. The protease treated silk was the exception, where no protein bands were detected throughout the entire lane. This anomalous behavior has been observed previously with electrophoretic separation of protease from *Streptomyces griseus* (Pronase E) and a number of other proteases. Autolysis of the enzyme during sample preparation was considered a possible cause. In the protease treated silk, there is a possibility of a preferential attack on the L-chain without scission of the silk main chain. This would be consistent with the loss of the 28 kDa band with no accompanying decrease in intrinsic viscosity and HPSEC molecular weights.

Figures 3 and 4 also show the presence of two additional molecular weight fragments at 17-18 and 10-12 kDa. They follow exactly the same behaviour with aging and treatment as the 28 kDa fragment and can be considered as characteristic of the fibroin degradation.

Usefulness of Viscometry, HPSEC, and SDS-PAGE

Both viscosity measurements and HPSEC showed that silk is stable in LiSCN, confirming that it is a good choice for a solvent. Viscometry results of Weather-Ometer aged silk confirmed that the technique is extremely sensitive to the early stages of photodegradation, and there is an excellent correlation to HPSEC. The sensitivity of both techniques decrease with further degradation. HPSEC provides additional information on the distribution of the polypeptides components over the molecular weight range. SDS-PAGE results provide information about the integrity of the silk L-chain with aging. The decrease in the 28 kDa band intensity with photodegradation is consistent with deterioration of the L-chain. This is also corroborated by the skewing of HPSE chromatograms. As the PAGE results are in agreement with viscometry and HPSEC, this shows that the 28 kDa band intensity can also be an indicator of the state of degradation of silk, but other techniques are required to corroborate the results.

As a routine method of analysis, HPSEC has several advantages over viscometry and electrophoresis. It requires simpler sample preparation and a smaller sample size, it is much less labour intensive, and gives information about the molecular weight distribution of silk. There are obvious limitations with all these techniques. First, they require polymers to be in solution. Analysis of weighted silk, for example, cannot be done unless the weighting is removed. Second, these methods are not suitable for detecting mechanical damage, end-wise degradation, or changes in surface texture. Third, for analyses of very degraded silk, these techniques are not very sensitive to small

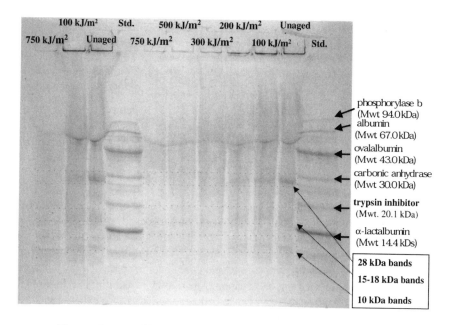

Figure 3   SDS-PAGE - Weather-Ometer Aged Silk

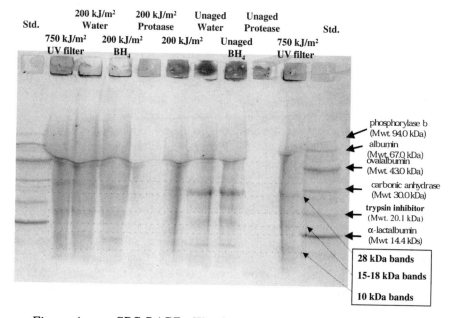

Figure 4   SDS-PAGE - Weather-Ometer Aged and Treated Silk

changes in molecular weight, but they can be used, if other well characterized reference samples are analyzed at the same time.

UV Filter Protection of Silk

It is well known that silk is susceptible to damage by light, especially the UV component of light. Because of this UV-filters are installed in most museums. The protection of a 400 nm cut-off UV filter for silk is clearly demonstrated by the results from all three techniques: a significantly higher intrinsic viscosity, higher HPSEC molecular weight averages and a higher intensity of the 28 kDa band in SDS-PAGE when compared with the unprotected silk. The level of protection provided by the UV-filter, at the intensity level used in the experiments, was estimated to be more than 7-fold based on viscometry and HPSEC results.

Although damage of silk by UV-light is well known, the effect of UV-filtered high intensity light is more uncertain. The question is often asked: is it safe to display silk textiles at higher light levels if a UV-filter is used? Both HPSEC and viscosity results showed that there is some degradation with UV-filter protected samples. The heat generated at higher intensity exposure would contribute to some of this, but it cannot account for all the degradation. We cannot rule out the possibility that high intensity light without a UV-component can cause damage to silk. More experiments will be carried out in the future to establish this.

Effect of Treatment on Light-aged Silk

One of the biggest concerns among textiles conservators is wet cleaning of degraded silk – the possibility that the treatments can further damage the textile. Water washing and the use of detergents are both acceptable treatments for new silk, but the long term effects on degraded textiles are still in question. Borohydride and protease enzymes have been successfully used by paper conservators as special stain removers. These are considered to be too risky for silk in good or bad conditions. With borohydride the concern is the possible reduction of disulphide bonds and the high pH (~10) of the aqueous solution. Proteases in general are hazardous for silk because of their ability to break peptide bonds. The Pronase E used in this study, for example, is a non-selective protease and is potentially able to cause a great deal of damage to silk. In this study, all four treatments were carried out on both unaged and light-aged silks in order to see if degraded silks are more susceptible to damage.

Both the viscosity and HPSEC results showed that there were no major decreases in molecular size of the silk main fibroin as a result of any of the treatments, and that the light-damaged silks were not more susceptible to damage from these treatments than unaged silk. But there were subtle differences among the treated and untreated silks for all treatments except water. The most obvious is the loss of the 28 kDa band in SDS-PAGE of the protease treated silk. Such loss of low molecular weight fractions can result in a relative increase in averaged molecular weight. This was confirmed by increase in $[\eta]$ accompanied with increased in M_p and decrease in M_n. Therefore the main fibroin is assumed to be intact. This loss of the 28 kDa band can either be the result of interaction between enzyme residue and the silk, interfering with detection, or that there is a selective destruction of the L-chain protein by the protease without measurable breakdown in the main fibroin. Like protease treatment, borohydride treated silk consistently have a higher [η] compared to controls. A reasonable explanation is that these treatments altered the silk conformation in solution making it less compact, which led to a larger hydrodynamic volume and a higher viscosity.

SDS treated silk, on the other hand, had consistently lower [η] than the untreated control. Degradation is an unlikely cause as more aggressive treatments like borohydride and protease did not cause measurable damage. A more plausible explanation lies in the fact that under denaturing conditions, SDS associates with proteins in solution and imposes a charge and a shape to the protein molecules. If this happens, to a very limited extent, between detergent residue and silk protein in solution, it could cause the silk to alter its conformation – making the silk protein more compact – resulting in a smaller hydrodynamic volume and a lower [η]. SDS-PAGE of borohydride and SDS treated silk showed very little difference in the 28 kDa band intensity, suggesting that these treatments do not affect the L-chain of the silk protein.

Conclusions

The intrinsic viscosity of unaged silk habutae in 6.67 M lithium thiocyanate was 0.40 dL/g at 30°C. Using globular protein standards, pH 7 tris-urea buffer as the mobile phase, the HPSEC calculated M_p for unaged silk was estimated to be 496 kDa, M_w was 346 kDa and M_n was 69.2 kDa. The SDS-PAGE band at approximately 28 kDa observed in unaged silk has been assigned to the L-chain subunit. Two additional low molecular weight fractions (17-18 and 10-12 kDa) were also found characteristic of the electrophoretic pattern of fibroin.

Viscometry using lithium thiocyanate solvent, has been confirmed to be sensitve to very early stages of photodegradation in *Bombyx mori* silk. Compared with viscometry, HPSEC is equally sensitive and more informative. There is an excellent correlation among the results from both techniques, both are very sensitive to changes in undegraded or moderately degraded silk, but much less sensitive with very degraded silk. SDS-PAGE provided additional information about the silk L-chain, such as its loss as a result of photodegradation and possibly the selective break down by proteases. Among the three techniques, HPSEC is most suitable as a routine technique because it is much less labour intensive, and requires a small sample size.

The effectiveness of UV-filters in protecting silk from light damage is also confirmed in the results. Among the treated silk, results showed that none of the four treatments, immersion in water, SDS, protease, or ethanolic solution of sodium borohydride, affect light damaged silk differently than undamaged silk. There is no evidence that these treatments cause damage to the main protein chain of the silk fibroin, although there is evidence suggesting that SDS residue may interact with silk in solution and protease may damage the light chain (28 kDa) of silk fibroin. All three techniques in this study will be used in future studies to evaluate the effects of more aggressive stain removers, cleaning techniques, and display environments on silk.

Acknowlegement

The authors would like to thank David Howell, conservation scientist at Hampton Court Palace, for giving us the detail procedure for HPSEC, for the many discussions and for his continuous encouragement during the development of this technique. We thank Lyndsie Selwyn for her ready advice, many excellent suggestions in the organization and clarification of the manuscript and painstaking efforts in making sure that nothing was missed, and Barbara Patterson for her thorough editing. To David Grattan, we owe our thanks for his continuous support and encouragement in this work and his very helpful suggestions in reviewing this paper. We also thank Maureen MacDonald for help with setting up the Weather-Ometer® and calibration of the xenon-arc lamp and Elizabeth Moffatt for helping us get started with the HPLC.

Literature Cited

1. Miller, J.E., Doctoral Dissertation, Kansas State University, KS, 1986
2. Becker, M.A., Doctoral Dissertation, Johns Hopkins University, MD, 1993
3. Hansen, E.; Sobel, H. *AIC Textile Specialty Group Postprints*; Thomassen-Krauss, S.; Eaton, L.; Reiter, S. L. Eds.; 20th Annual Meeting, Buffalo, N.Y. June 1992, Vol.2, pp. 14-30
4. Kurupillai, R.V.; Hersh, S.P.; Tucker, P.A., *Historic Textile and Paper Materials Conservation and Characterization*, Needles, H.L.; Zeronian, S.H., Eds; Advances in Chemistry Series No. 212; ACS: Washington, DC., 1986; pp.111-130
5. Hansen, E.F.; Ginell, W.S. *Historic Textile and Paper Materials II Conservation and Characterization*, Needles, H.L.; Zeronian, S.H., Eds; ACS Symposium Series No. 410; ACS: Washington, DC., 1989; pp.109-133
6. Halvorson, B.G.; Kerr, N. *Studies in Conservation* **1994**, *39*, pp. 45-56
7. Miller, J.E.; Reagan, B.M. *J.AIC* **1989**, *28*, no. 2, pp. 97-115
8. Van Oosten, T.B. *The Degradation of Fibroin Under the Influence of Weighting - Present State of Knowledge*, Internal Report from Centraal Laboratorium van Kunsten Wetenschap - Central Laboratory for Art and Science, Amsterdam, 1992
9. Lemiski, S.L. Master Thesis, University of Alberta, Alberta, Canada, 1996
10. Tímár-Balázsy, Á; Mátéfy, G.; Csáányi, S. *ICOM Committee for Conservation 10th Triennial Meeting Preprints, Washington, DC, August 22-27, 1993*, Bridgland, J. Ed.; ICOM Committee for Conservation: LA, 1993, Vol. 2, p. 330

11. Shashoua, Y. *ICOM Committee for Conservation 9th Triennial Meeting, Dresden, GDR, August 26-31, 1990: preprints*; Bridgland, J. Ed., ICOM Committee for Conservation: LA, 1990, Vol. 2, p.313

12. Ballard, M.; Rhee H. *ICOM Committee for Conservation 10th Triennial Meeting, Washington, DC, August 22-27, 1993: preprints*; Bridgland, J. Ed.; ICOM Committee for Conservation: LA, 1993, Vol. 2, p. 327

13. Yau, W. W.; Kirkland, J. J.; Bly, D. D. *Modern Size-Exclusion Liquid Chromatography Practice of Gel Permeation and Gel Filtration Chromatography*, John Wiley & Sons, Inc., NY, 1979

14. Browning, B.L. *Methods of Wood Chemistry*, Interscience Publisher: N.Y., 1967, Vol.2, pp. 519-557

15. Tweedie, A.S. *Canadian Journal of Research*, **1938**, *16 Section B*, p. 134

16. Barth, H.G., Boyes, B.E. *Anal. Chem.*, **1992**, *64*, pp. 428R-442R

17. Harris, D.A. In *Methods in Molecular Biology: Practical Protein Chromatography*, Kenney, A.; Fowell, S., Eds; The Humana Press Inc., NJ, 1992, Vol.11 pp. 223-236

18. Chicz, R.M.; Regnier, F.E. *Methods in Enzymology: Guide to Protein Purification*, Deutscher, M.P. Ed.; Academic Press, Inc.: NY, 1990, Vol. 182, pp. 392-421

19. Le Maire, M.; Ghazi, A.; Møller, J.V. *Strategies in Size Exclusion Chromatography*; Potschka, M.; Dubin, P.L. Eds; ACS Symposium Series No. 635; ACS: Washington, DC., 1996; p.37

20. Sasaki, T.; Noda, H. *Biochim. Biophys. Acta*, **1973**, *310*, pp. 77-90

21. Sasaki, T.; Noda, H. *Biochim. Biophys. Acta*, **1973**, *310*, pp. 91-103

22. Howell, D. In *Silk*, Harpers Ferry Regional Textile Group Conference Preprints; National Museum of American History; November 12-13, 1992, pp.11-12

23. Deyl, Z. In *Electrophoresis a Survey of Techniques and Applications, Part A: Techniques*; Deyl, Z., Everaerts; F.M., Prusik, Z.; Svendsen, P.J., Eds.; Journal of Chromatography Library; Elsevier Scientific Publishing Company Amsterdam, 1979, Vol. 18, pp.45-67

24. Trefiletti, V.; Conio, G.; Pioli, F. Cavazza, B. *Makromol. Chem.*, **1980**, *181*, no.6, pp. 1159-1179

25. Oyama, F.; Mizuno, S., Shimura, K. *J. Biochem.*, **1984**, *96*, pp. 1689-1694

26. Kodrík, D. *Acta Entomol. Bohemoslov.*, **1992**, *89*, pp. 269-273

27. AATCC Test Method 16-1990, "Colorfastness to Light , Option E: Water-Cooled Xenon-Arc Lamp, Continuous Light"

28. Lucas, J.; Shaw J. T. B.; Smith, S. G. In *The Silk Fibroin*; Advances in Protein Chemistry, 1957, Vol. 13, pp. 107-242

29. Otterburn, M.S. *Chemistry of Natural Protein Fibres*; Asquith R.S. Ed.; Plenum Press: NY, 1977, p.53

30. Waldschmidt-Leitz, E.; Zeiss, O. *Z. Physiol. Chem.*, **1955**, *300*, p. 49

31. Signer, R.; Strässle, R. *Helv. Chim. Acta*, **1947**, *30*, p. 155

32. "Textiles: Testing for Fibre Alteration and Fibre Damage: Determination of the Viscosity Number of Natural Silk in Lithium Bromide Solution" *Standard, Switzerland* SNV-19559, 1968

33. Garrett, C. H.; Howitt, F. O. *Journal of the Textile Institute*, **1941**, *January*, pp. T1-T12

34. Billmeyer, F.W. Jr. *Textbook of Polymer Science*, 3rd ed., J. Wiley & Sons: N.Y, 1984, p.208

35. Schulz, G.V; Blaschke, F. *J. Prakt. Chem.*, **1941**, *158*, p. 140

36. Chromatographic Specialties, Inc., PO Bag 1150, Brockville, ON, Canada, K6V 5W1

37. The procedure used in this study was modified from that described by Howell - personal communication with David Howell, conservation scientist, Hampton Court Palace, 1997.

38. Peters R.H. In *Textile chemistry*, Elsevier, 1963, Vol. 1, p.309

39. Tashiro, Y.; Otsuki, E.; Shimadau, T. *Biochim. Biophys. Acta*, **1972**, *257*, p. 198

40. Schade, W.; Liesenfeld, I.; Ziegler, Kl. *Kolloid-Z. U. Z. Polymere*, **1970**, *242*, pp. 1161-1164

41. Song, Z-T.; Speakman, P.T. *Biochem. Soc. Transactions*, **1985**, *13*, no. 2, p.387

42. Kaplan, D.; Adams, W.W.; Farmer, B.; Viney, C. In *Silk Polymers-Materials Science and Biotechnology*; Kaplan, D.; Adams, W.W.; Farmer, B.; Viney, C Eds; ACS Symposium Series No. 544; ACS: Washington, DC., 1994; pp. 2-16

43. Becker, M.A;, Tuross, N. In *Silk Polymers-Materials Science and Biotechnology*; Kaplan, D.; Adams, W.W.; Farmer, B.; Viney, C Eds; ACS Symposium Series No. 544; ACS: Washington, DC. ., 1994; pp. 252-269

44. Kerr, N. In *Silk,* Harpers Ferry Regional Textile Group Conference Preprints; National Museum of American History; November 12-13, 1992, p.5

Chapter 9

The Aging of Wool Fibers

Ian L. Weatherall

Cavendish Scientific and Technical Consultants Ltd., P.O. Box 1189, Dunedin, New Zealand

The aging of wool fibers results in their degradation. In essentially all circumstances the main chemical changes involve the effects of light, heat, and atmospheric oxygen. These changes begin from the time the sheep is born. Their subsequent progression under a variety of natural and artificial conditions has been studied. Modifications in physical properties may be related to the underlying chemical changes. Common features of the effect of both light and heat are protein chain scission with formation of keto-acyl end groups, cross-link cleavage, and also their formation. Light stable cross-link formation, possibly in the form of hemithioacetals, appears particularly facile. Main chain scission in combination with cross-link formation ultimately leads to the severe embrittlement of wool fibers.

The Morphology and Chemistry of Wool.

Wool fibers have average diameters ranging from about 20-40 microns. Within any group of fibers there is considerable variation in the average diameter. Along the length of an individual fiber the diameter may vary by a few microns. The typical appearance of a wool fiber observed with a scanning electron microscope is shown in Figure 1. The fiber is covered by scales, which are collectively known as the cuticle. These scales are flattened, overlapping cells, about 0.5 microns thick, with the scale edges pointing towards the tip of the fiber. The main body of the fiber is called the cortex, which makes up about 90% of the whole. The cortex consists of elongated, spindle-shaped cells, about 100 microns long, and 4-5 microns thick at their widest part. Cell membranes about 0.25 microns thick attach the cortical cells to each other, and the scale cells to the underlying cortex. A diagrammatic representation of these main morphological features within a wool fiber is shown in Figure 2.

Figure 1. Scanning Electron Micrograph of a Wool Fiber

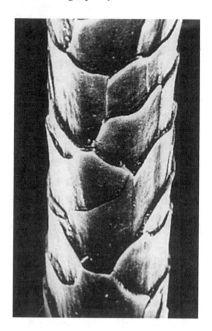

Figure 2. Diagrammatic Representation of the Morphology of a Wool Fiber

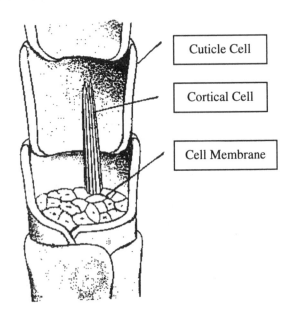

Cuticle Cell

Cortical Cell

Cell Membrane

The mechanical properties of the wool fiber are essentially those of the cortical cells [1]. These have a substructure consisting of filamentous material called macrofibrils, which are about 0.3 microns thick, and span the length of the cells. The macrofibrils are made up of smaller units termed microfibrils which have been regarded as the basic mechanical units of the wool fiber. The microfibrils consist of axially oriented, ordered assemblies of helical protein chains, associated with some amorphous proteins. Together with other constituents, these proteins constitute the basic molecular structures of wool, in the form of specific sequences of some twenty-two different amino acids. They are known collectively as keratin proteins.

Wool keratin protein contains many disulfide bonds. These cross-link the protein chains, and the mechanics of the fiber may be regarded as those of a cross-linked polymer network [1-3]. More than half of the protein is fibrillar, of relatively high molar mass, and oriented along the fiber axis. Within the fibrillar proteins about one amino acid residue in fifteen consists of a half-cystine residue. The fibrillar proteins have a significant influence on the mechanical properties of wool. They are associated with a matrix of other proteins of lower molar mass, and with a higher content of disulfide bond cross-links. In any study of wool fiber aging the integrity of the main protein chains and the inter-chain cross-links are the chemical entities of interest. The tensile strength of the dry wool fiber depends largely on the length of the main protein chains and not on the covalent cross-links [2]. The dry strength has been reported to decrease by about 25% after scission of 2.5% of the peptide bonds. Fiber strength is halved with 5% main chain scission. In comparison, fiber weakening only occurs after about 60% of the cross-links have been broken. Deleterious changes to wool may be caused by heat. However, at ambient temperatures thermal degradation is negligible, except over very long time periods [4]. In contrast, the effect of light on wool is particularly damaging, and can occur over short periods of time. The general nature of the chemical changes is illustrated in Figure 3.

Figure 3. Photochemical Changes which Affect the Mechanical Properties of Wool.

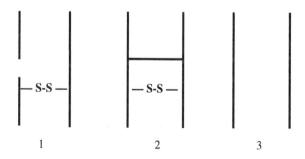

(1) Scission of the main protein chain
(2) Formation of new cross-links
(3) Destruction of existing cross-links

The chemical and related physical changes include those which occur first under natural conditions during growth, then those which occur during the time the fibers are in textile use, and finally any subsequent very long term changes. The last of these has probably been the least studied, but in the context of historical textiles they may be particularly important, since keratinous substances housed in museum environments have been reported to undergo further degradation [5].

Aging During Growth.

Change to the characteristics of wool during growth was recorded over 100 years ago [6]. Subsequent early reports noted chemical and physical changes caused by weathering [7]. A more recent study reported that fibers sampled from lambs born under darkened indoor conditions were significantly different from those from lambs born outdoors [8]. The former showed higher solubility following performic acid treatment and in urea-bisulfite solutions. These measurements reflect the level of protein cross-linking not susceptible to either oxidative or reductive cleavage such as disulfide bonds. The lower solubilities of wool from lambs born outdoors suggested that it contained some new form of protein of cross-linkage. In the same report fibers sampled from the sheep born and maintained under darkened conditions showed no end-to-end chemical or physical differences [9]. In contrast, fibers sampled from sheep born and maintained under natural outdoor conditions showed significant end-to-end differences. The tip halves had lower performic acid and urea-bisulphite solubilities, again indicative of the formation of new protein cross-linkages. Natural light exposure of the fibers was the logical causative agent.

Aging During Textile Use.

Wool fibers at the beginning of their textile use will have been already subjected to a variety chemical and mechanical influences during manufacturing processes. Despite this the aging of wool fibers might be considered to begin during their use in textile products. This is reasonable, since the actual practical use of wool textiles can only be considered to start after the fibers have been subjected to manufacturing processes. However, from a more considered scientific viewpoint, all factors need to be included within the continuum of events which make up the entire aging process.

The influences on wool during both manufacture and use are many. The most common are light, heat, water and mechanical action. All these need to be considered, as well as any others within the context of long term aging. The effect of light has received considerable attention from wool scientists [2,3]. This paper will address some of the more recently reported results.

Wool fibers which had been subjected to minimal processing, were exposed progressively to simulated sunlight in the form of an MBTL lamp. This light source has been shown to produce the same effects as sunlight [10,11], and has been used to examine the degradation of keratinous materials in a museum environment [12]. The effect on single fiber strength is shown in Figure 4.

Figure 4. Reduction of Wool Fiber Breaking Strength on Irradiation.

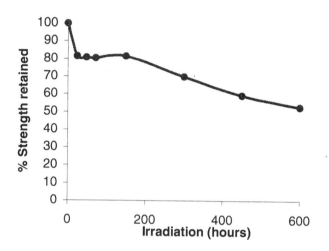

The strength decreased rapidly during an initial short exposure period. It then changed little with time before a longer term decrease. The strength loss was indicative of main protein chain scission. The dry strength of wool is known to be sensitive to breaks in the protein chain [2,3]. Light was the obvious causative agent.

The scission mechanism induced by light and oxygen was first demonstrated for simple amides [13]. For peptides in the solid state the same light induced scission mechanism has been shown to occur [14,]. Therefore, the nature of protein chain scission in aged wool appears to be well established. The primary chemical event is the homolytic scission of the C-H bond at the alpha carbon of an anhydro amino acid residue along the protein chain. This results in the formation of a free radical on that carbon. The kinetics of its formation in wool irradiated with blue light in a 10 nm waveband centered on 420 nm has been studied in some detail [15,16]. It has been reported to form in wool as a result of raising the temperature of the protein [17], and has been reported in wool as a result of simple mechanical action on the fiber [18]. A reaction of this radical with a molecule of oxygen would lead ultimately to protein chain scission with formation of both primary amide and keto-acyl end groups. This sequence is shown in Figure 5.

Figure 5. Sequence of Protein Chain Scission.

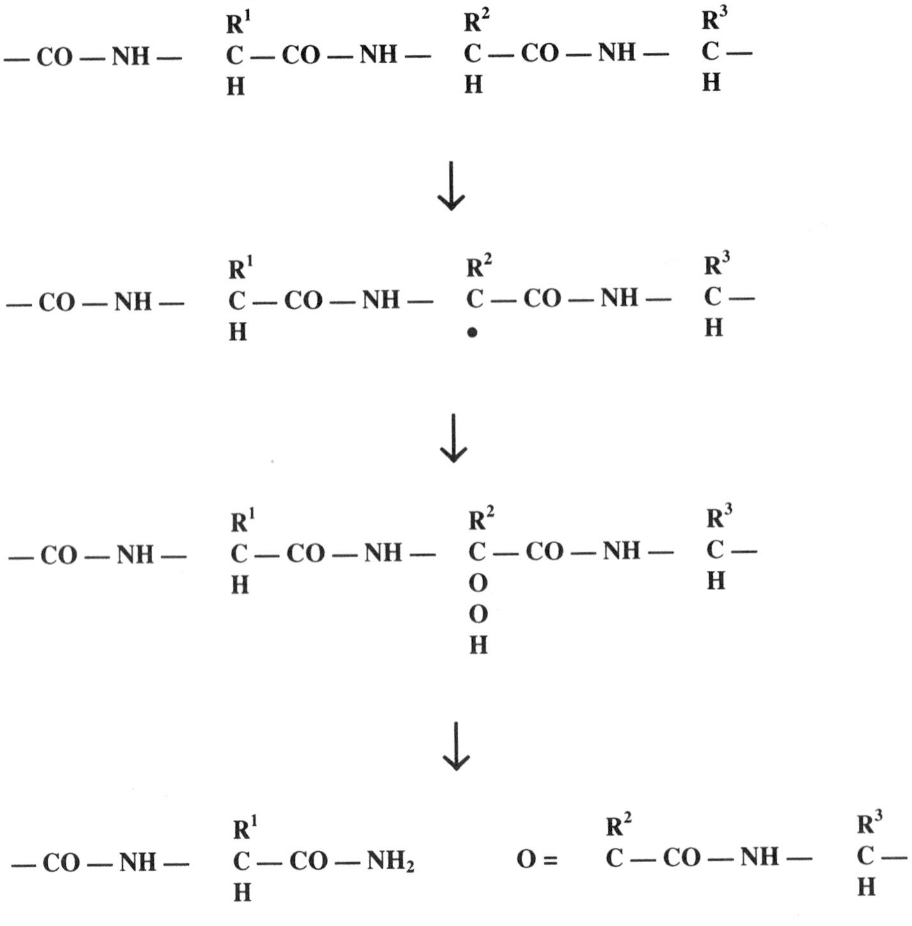

Analysis of the levels of both light induced end groups has been reported for light degraded wool [19,20]. In the case of keto-acyl end groups, the total of all the individual end groups has been shown to increase with the time light exposure. These levels are shown in Figure 6.

The rise in keto-acyl end groups in wool irradiated with simulated sunlight provided evidence that the mechanism of amide scission demonstrated for simple amides [13], and for peptides in the solid state [14], also occurred in wool proteins.

Figure 6. Total Keto-acyl Content of Light Degraded Wool.

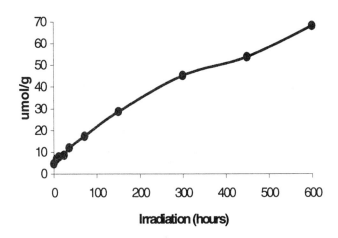

The performic acid solubility of wool has been reported to change significantly after exposure to simulated sunlight [21]. The effect is shown in Figure 7. There was a significant decrease after short exposure, and after the initial minimum began to increase with subsequent exposure. The initial decreased solubility implied the formation of protein chain cross-linkages, while the longer term increase could be accounted for by concurrent main chain scission.

Figure 7. Light Induced Changes in the Performic Acid Solubility of Wool

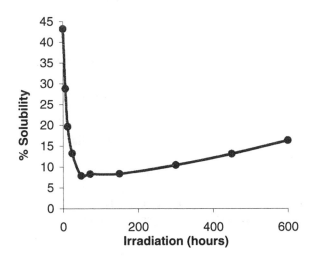

It has been by shown by chemical analysis of wool following progressive exposure that there was a slow overall loss of the constituent keratin protein amino acids. The most significant loss was that of cystine [20] as shown in Figure 8.

Figure 8. Loss of Wool half-cystine Content.

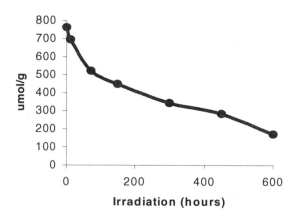

However, elemental analysis of exposed fibers has been reported to show no significant overall loss of sulfur [7]. The products of wool cystine photo-degradation have long been of interest. Photo-reduction of the S-S bond to produce cysteine has not been demonstrated, even though light exposed wools appear to give strong positive reactions for the presence of thiols [3,7]. Furthermore, only a small proportion of the lost cystine could be accounted for as cysteic acid [20]. This indicated that photo-oxidative reaction products of the cystine sulfur were not those initially formed. One other possibility has been proposed. Namely the formation of perthiol groups [25]. An analytical method for perthiols was subsequently developed [26], and when applied to light exposed wool showed that it appeared to contain this functional group [21].

In the case of heat degraded wool it has been shown that there was an essentially linear relationship between the loss of strength and the rise in keto-acyl end groups [27]. For light degraded wool the relationship was not linear, and this may be a consequence of the observed concurrent light induced formation of protein cross-links.

It has been proposed by Weatherall and France that these cross-links include the product of reaction of the of perthiol and keto-acyl groups [21]. The reaction of thiols, and presumably perthiols, is a well known reaction leading to thiohemiacetal formation. However, it has also been established that irradiated wool contains "apparent" thiol [3]. Therefore, it has been proposed that the perthiohemiacetals rearrange to give the thiohemiacetals of thioketo-acyl groups as follows:

$$R\text{-SSH} + O\!=\!C(R)\text{-CO}... \longrightarrow RSSC(OH)(R)\text{-CO}... \longrightarrow RSC(SH)(R)\text{-CO}...$$

The formation of this crosslink functionality would explain many of the observed chemical properties of irradiated wool. In particular it would explain, at least in part, some of the outstanding questions about the fate of cystine following irradiation. Its formation involves no loss of sulfur, and it contains a reactive thiol group. Such a crosslink could exist in both intra - and inter - molecular forms, and its formation would depend on the relative rates of formation of appropriately positioned perthiol and keto-acyl groups. Molecular and computer models of a perthiol group within an alpha helical protein chain have shown that it would be appropriately positioned to react with a keto-acyl group four amino acid residues along the chain. Such an intra-chain reaction would constitute a self repair mechanism for light induced wool keratin protein chain scission. The reaction between protein chains would form a crosslink in the normal polymer sense. Light induced changes to the crosslink levels in wool are consistent with observed changes in solubility and swelling [8], and other mechanical properties [9].

Long Term Aging.

There appears to be a paucity of published reports of details of the chemical and physical characteristics of wool fibers long after the end of their textile use. Such samples as may exist, for example in museums, will have had a diverse history. The influences to which they may have been subjected will have been many and varied. Some of them may have been subjected to little more than mechanical action during their history. Others may show the result of many influences, including light, heat, and moisture. One characteristic which may reflect the relative influence of past agents, is the fracture morphology of individual fibers as revealed by scanning electron microscopy. The normal fracture morphology of wool shows features related to its complex internal structure at the cellular and subcellular level. Exposure of wool to simulated sunlight for periods of time from 24 to 240 hours has been shown to result in progressive changes to the mode of fiber fracture [28]. For wool fibers subjected to long term exposure to light, a smooth, transverse fracture morphology has been reported, indicative of a highly cross-linked and/or weakened interior molecular structure [22,30]. Therefore, in wool subject to long term aging the examination of this characteristic, in combination with an analysis of the chemical and physical properties discussed above, could provide information not only about its past, but also be a guide to the appropriate very long term conservation environment. A particularly noteworthy study of very old keratin fibers has been reported [29]. Fiber strength and fracture morphology by SEM has been reported for Egyptian Mummy hairs from sites dated at 1500-4000 years old. The fibers had about 10-20% of the strength of modern hair, but retained less than 10% extensibility. Of particular significance was the fracture morphology of the very old fibers. The fracture morphology of new and moderately aged wool usually shows details characteristic of the complex morphology at the cellular and subcellular level as shown in Figures 9 and 10. In contrast, the fracture surface of severely aged wool has been reported to be smooth and relatively devoid of detail [30]. This characteristic is shown in Figure 11.

Figure 9. Scanning Electron Micrograph of a Broken New Wool Fiber.

Figure 10. Scanning Electron Micrograph of Wool Fiber Exposed to 1000 hours Simulated Sunlight.

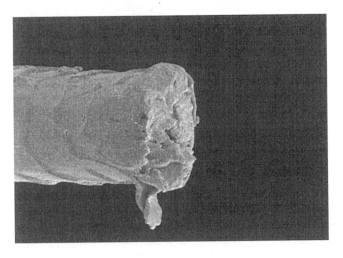

Figure 11. Scanning Electron Micrograph of Severely Photo-degraded Wool Fiber.

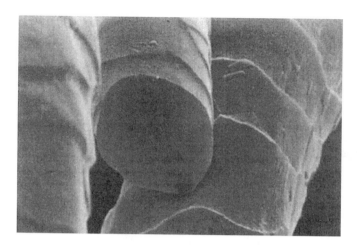

The severely aged wool in this SEM was exposed to glass filtered natural sunlight for 2500 hours.

Acknowledgements.

The assistance of Dr Peter Cooke and Mr Douglas Soroka with SEM (Figures 9 and 10), and Dr Fenella France in the preparation of this article, is gratefully acknowledged.

References.

1. R. Postle, G. A. Carnaby, and S. de Jong, *The Mechanics of Wool Structures*, Ellis Horwood, Chichester, 1988.
2. P. Alexander and R. F. Hudson, *Wool: Its Chemistry and Physics*, Reinhold Publishing Corporation, New York, 1954.
3. J. A. Maclaren and B. Milligan. *Wool Science*, Science Press, Marrickville, NSW, 1981.
4. Werner von Bergen. *American Wool Handbook*, Interscience Publishers, N.Y. 1963.
5. C.V. Horie, *Polym. Deg. Stab.*, 29 109 (1990).
6. Löbner, *Flecken in Wollwaren*, 134 (1890).
7. J. B. Speakman and P.R. McMahon, *N. Z. J. Sci.Tech.*, 1B, 32; 5B, 248, (1939).
8. I. L. Weatherall and L. A. Dunn, *Photochem. Photobiol.*, 55(2) 305 (1992).
9. L. A. Dunn and I. L. Weatherall, *J. Appl. Polym. Sci.*, 44, 1275, (1992).

10. J. Park and D. J. Smith, *J. S. D. C., 90(12) 431, (1974)*.
11. I. H. Leaver, R. C. Marshall, and D. J. Rivett, in *Proc. 7th Int. Wool Text. Res. Conf.*, Tokyo, IV, 13, (1985).
12. C.V. Horie, in *Proc. 9th Triennial Meeting of the International Council of Museums*, Dresden, 431 (1990).
13. W. H. Sharkey and W. E. Mochel, *J.A.C.S.*, 81, 3000 (1959).
14. A. Maybeck and J. Maybeck, *Photochem. Photobiol.*, 11, 53, (1970).
15. R. F. C. Claridge, A. Shatkay and I. L. Weatherall, *J. Text. Inst.* 7, 316, (1979).
16. A. Shatkay and I. L. Weatherall, *J. Polym. Sci., 15, 1735, (1977)*.
17. A. Maybeck and J. Maybeck, *Applied Polymer Symposia*, 18, 325, (1971).
18 J. J. Windle and A. K. Wierema, *J. Appl. Polym. Sci.* 8, 1531, (1964).
19 I. L. Weatherall, *The Tendering of Wool by Light*, New Zealand Wool Research Organisation, Research Report No.22, 1973.
20 F. G. France and I. L. Weatherall, In *Proc. 9th Int. Wool Text. Res. Conf.*, Biella, II, 100, (1995).
21 I. L. Weatherall and F. G. France, In *Proc. 3rd Asian Text. Conf.*, Hong Kong, I, 892, (1995).
22. I. L. Weatherall, In *Proc. 5th Int. Wool Text. Res. Conf.*, Aachen, Sonderband der Schriftenreihe, 2, 580, (1975).
23. L. A. Dunn and I. L. Weatherall, In *Proc. 8th Int. Wool Text. Res. Conf.*, Christchurch, I, 619, (1990).
24. F. G. France and I. L. Weatherall, In *Proc. 8th Int. Wool Text. Res. Conf.*, Christchurch, I, 609, (1990).
25. B. J. Sweetman, *Text. Res. J.* 36, 718, (1966).
26. D. Calvini, G. Federici and E. Barboni, *Eur. J. Biochem.*, 14,169, (1970).
27. L. Lu and I. L. Weatherall, in *Proc. 9th Int. Wool Text. Res. Conf.*, Biella, II, 265, (1995).
28 M. Zimmermann and H. Höcker, *Text. Res. J.*, 66, 657, (1996).
29 E. R. Massa, M. Masali, and A. M. C. Fuhrman, *J. Human Evolution*, 9,133, (1980).
30 N. S. Allen, M. Edge, and C. V. Horie (Eds), *Polymers in Conservation, Special Publication No.105*, The Royal Society of Chemistry, Cambridge, 1992.

Polymers in Museums

Chapter 10

Polymers in Museums

Mary T. Baker

Egypt MVE, 1133 20th Street, N.W., Suite 600, Washington, DC 20036

This chapter reviews the importance and historical significance of polymers in museums. Conservation research for polymer based artifacts are discussed including the identification of unknown polymers, the assessment of current deterioration and the prediction of future deterioration of polymers, proper storage and display conditions, and new test methods for the preservation of museum objects.

Polymers have been in museums for as long as there have been museums. Natural polymers, modified or not, are ubiquitous in a museum, either as part of an artifact or as part of the repair and care of artifacts. With the advent of synthetic polymers in the late 19th century, the number and types of polymers found in a museum increased, along with the number and types of associated problems. Addressing these problems is certainly not a new activity and no material in a museum is expected to last forever. However, changing perceptions of the importance of artifacts containing polymers (particularly from the synthetic polymer era) have caused raised awareness of their often ephemeral nature.

The "cheap plastic" of the 50s and 60s is recognized by the museum-going public as an important part of our cultural history. Even though many of these materials are still in use, the importance of keeping well-preserved examples for future generations is apparent. These objects were originally considered more durable and therefore less in need of special attention than older materials; experience has shown that "durable" in everyday usage does not mean "neglectable" in a museum. By extension, many museum professionals recognize that modern acquisitions from later decades must have proper attention from the start, based on the recommendations from conservation scientists.

Conservation research on polymeric materials can be grouped into categories:

- identification (of unknowns, and determination of specific composition – "reverse manufacturing")
- assessing the current state of deterioration and predicting future deterioration,
- determining proper storage/display conditions,
- testing new polymers for use in preserving other works of art.

Scientists working in museums or in collaboration with museum professionals continue to advance all of these areas, both by expanding the group of materials studied (including new materials made available by advances in polymer synthesis) and by exploiting advances in instrumentation to increase accuracy, reduce sample size, and interpret information from polymers in museums.

Identification

What Is It Made Of?

The simple question, "what is it made of?" is probably asked about every artifact; the answer is sometimes not self-evident, or sometimes more certainty is considered necessary. The history of the piece may be in doubt, the information may be needed for a study of the artisan, there may be later additions (repairs or spills), or the identity of the materials may be needed for proper collections management. Unfortunately, complete identifications of all the materials in all objects can not be done – mostly for cost reasons – so such a justification for analysis must be made; the curiosity of a curator or scientist is generally not a sufficient reason.

As with most things, "what is it" can have many levels as a question, so the answer can be given on many levels. It may be sufficient to determine whether the materials are natural or synthetic polymers, or it may be necessary to know everything about the polymer: identification of repeat unit(s), molecular weight, molecular weight distribution, branching, crosslinking, etc. In the first case, an answer may be easily found with minimal sample and low-tech methods; in the other extreme, even with larger samples and a generous analytical budget, the complete answer might be elusive.

One aspect of identification that needs work is standards. A chemist might be asked to identify the polymeric component of a painting, a piece of sculpture, a kitchen utensil, a spacecraft part, a toy, jewelry, an archaeological artifact; the list could go on. A set of standard pure polymers can serve as a base for comparison for many of the modern synthetics, but can be inadequate if the artifact has undergone sufficient aging, introducing differences caused by oxidation, hydrolysis and inter-chain reactions. In the case of natural resins, modified natural polymers (such as cellulose nitrate), proteins and alkyds, precise matching standards are unlikely to be found.

Instrumental analysis, whether used to compare a sample to standards, or as an independent means of identification, follows standard laboratory practices, with variations to allow for smaller sample sizes or unusual compositions. Fourier Transform Infrared spectroscopy can generally characterize the polymer and often can be used to identify it; very small samples can be used in FTIRs with microscope attachments. Other spectroscopic methods, such as UV-Vis and Raman, can be used to further characterize the artifact, but are not usually helpful in identification. Chromatography is usually a next step: GC/MS is very useful, although the additional complication of derivatization is usually necessary, for natural resins and oils. For synthetic polymers, a pyrolysis attachment to a GC/MS can yield identifications, based on the monomers and degradation products. Liquid chromatography, particularly when coupled with an FTIR, can be used to separate and identify the soluble components of most of the polymers found in museums; its main disadvantage is its need for larger sample sizes.

Physical methods, such as mechanical testing and microscopy are useful in characterization, but rarely can be used for identification. Chemical tests abound for polymers, and can be used to make a precise identification in many cases. For qualitative tests, the sample size can be quite small, but for quantitative tests, such as determining the amount of aromatic groups in an alkyd, larger sample sizes are usually necessary.

Identification can be further complicated by the age and composition of the artifact: age-related changes in the polymer, degradation products, and fillers and other additives can all interfere with any identification method, and make it difficult to set standard analysis methods, such as those used for modern polymers. The conservation scientist depends on experience with similar materials to help choose the best analytical paths for an identification; with such a variety of polymers and artifacts, as well as compositions and conditions, it is more usual that any new sample will be "new ground."

What Else Is In There?

Many commercial laboratories offer reverse manufacturing as a service; with sufficient sample and money, one can learn all the components of a new product, and infer the likely synthetic and manufacturing routes. This is precisely the information that curators often want about a plastic sculpture or a historic piece of house ware or the paint in a modern painting, etc. Obviously, the manufacturing of the artifact is important; without that information, the artifact can only tell a partial story. Knowing how, when and with what the artifact was manufactured helps the curator develop a robust history around that particular piece, using it as an example of the advances and trends in the materials of its time.

Therefore, knowing that a piece of sculpture is cellulose nitrate is minimally useful, but knowing the method of nitration, source of cellulose, degree of nitration, type and amount of plasticizer and other additives and processing methods makes the object a record of that area of plastic manufacturing. It may help to place the artist's source of materials – for example, the artist might have been dissatisfied with the plastic sheeting sold to artists at the time and bought his materials from an airplane manufacturer.

Similarly, a polyolefin kitchen bowl set might seem to be one of many similar ones, but analysis reveals that polyethylene was used for some parts while polypropylene for others. Combined with manufacturing records of the time, this information can lead to a fuller understanding of how polymers were chosen for different products at the time; the set might be a one-of-a-kind oddity or it might reflect a decision made by the manufacturer, based on the desired properties.

Of all polymers in museums, modern paints are the most likely to depart from any expected or known compositions. Paints based on the early synthetics, such as acrylics and alkyds, were likely to have been mixed up in small batches, with variations from batch to batch. Artists often obtained the plain polymers, solvents and pigments and mixed their own, adding other materials to get certain effects. A fair amount of experimentation occurred when water-born acrylics became available; even with water-born acrylic paints being manufactured today on a larger scale, they are made in batches, with the components being changed to allow for differences in the pigments. As modern artists do not always keep records of the materials used and any post-production additions they made (and some records have been found to be faulty, either because of poor memories or misinformation) determining the complete composition of modern paints can be important to a technical study of an artist's work. Unfortunately, it is often paintings that yield the smallest samples, making a complete study difficult.

The technical study of Siqueiros' paintings by Contreras Maya, et al., highlights the importance of careful analytical analysis. As an artist whose career spanned the transition from oil paints to acrylics, Siqueiros and his choices in media during that time of great interest to art historians. Contreras Maya found that paintings originally designated as cellulose nitrate were actually done in other media – some oil and one acrylic. The misattributions were based on visual characteristics, partly following an assumption that cellulose nitrate media aged rapidly. Instead, she showed that the poor appearance of the paintings was due more to his unorthodox painting methods, rather than the failure of the polymer. Further work such as this is necessary to prevent the perpetuation of similar misconceptions.

Deterioration

Deterioration is an inevitable part of polymers in museums, and it is often essential to either assess the current state of deterioration or predict the course of future deterioration. Once again, this is an area that has a solid body of research in the non-museum world from which to draw. Determining amount and cause of degradation in industrial polymers is essential to quality control and maintenance, and predicting service lifetime is critical to almost all polymer products. However, such industrial research (usually) has the advantage of large samples and series of identical materials exposed to different conditions. It often relies on artificial aging, which requires a knowledge of all the components of the products and their aging properties (and interactions) to be useful. These conditions are rarely found in polymers in museums.

132

Assessing Current Deterioration

Knowing the current state of deterioration of a polymeric artifact helps the curator assess its age and history and helps determine whether the artifact should be included in a collection, how it is stored and whether it is exhibited. Evidence of old technology and early materials (such as reaction-molded phenolics) do not necessarily mean that a piece is old – methods and materials are revived for both economic and aesthetic reasons. Additionally, also for aesthetic reasons, a piece might be artificially aged. Lastly, differentiating between deterioration caused by neglect and that caused by use in service is important; displaying an object without discussing these aspects gives the visitor a false image of polymeric materials.

Well-documented objects can serve as standards of a sort: one advantage to working with polymers in museums is that there are often large collections of similar objects that have remained at constant temperature, relative humidity and light levels for a known number of years. This can provide very good examples of aged polymers without the uncertainty of artificial aging.

The work by Connors, et al., shows how this advantage can be exploited. By comparing artificially aged vulcanized natural rubber samples to similar samples taken from museum objects, Connors produced some typical landmarks in the FTIR spectra to use as a guide in evaluating unknown rubber samples. She also compared several different FTIR techniques, evaluating their usefulness in analyzing aging rubber samples. The results reflected some of the problems in working with older samples: in some cases a good formulation match did not seem to be made between museum sample and standard, and sensitivity of the analytical techniques decreased with age of the sample, mostly due to changing baseline and broadening of peaks. She recommends further research using the most promising FTIR technique and aging regimen that had resulted in the most similar results to naturally aged samples.

Future Deterioration

Discovering the history of an artifact helps flesh out and record the associated technology and determining the amount of deterioration present is fundamental to the artifact's history. However, predicting the future of the artifact is also critical; all the information contained in the artifact might otherwise be unexpectedly lost. For example, a poly (vinyl chloride) object that has just begun its autocatalytic deterioration stage in which the deterioration rate increases dramatically must be made a priority for study and preservation, as every day will matter. On the other hand, a vulcanized rubber object that has reached the end of its rapid degradation and, although in bad shape, is not going to change much if kept in a good environment can be scheduled for later with little loss of information.

General guidelines for predicting the future deterioration of a piece are gleaned from aging studies on similar polymers and plastics; obviously such guidelines would have the same exceptions as those noted for determining present states of deterioration. Some polymers have seemed quite unpredictable: for example, cellulose nitrate objects that were made more or less at the same time in the same factory, and kept in the same conditions, could not be counted on to deteriorate at the same rate. This suggested that small differences in synthesis and processing could cause great differences in the long-term aging properties of the polymer.

Ballany, et al., investigated the cause of these differences in cellulose acetate artifacts. While previous work on cellulose nitrate had indicated a relationship between residual acids from synthesis and an acceleration of aging, the work on cellulose acetate showed no similar relationship. However, the differing rates of deacetylation can be measured on objects using the non-destructive methods devised by Ballany, and the future deterioration rate predicted. Continued work may reveal root causes for differences in deacetylation rates.

Work such as Connors' and Ballany's is essential to preserving polymers in museums. Without instrumental methods to monitor deterioration of artifacts, the curator must rely on visual, olefactory and tactile observation. By the time deterioration is sufficiently high to be noticed by these methods, the object is very far down the aging path and it is often too late for any preservation attempts. Also, studies like theirs, using instrumental analysis, establish non-subjective baselines to which the artifacts can be compared over time, and may eventually yield more information as to some of the more subtle mechanisms of aging.

Storage and Display

Conservation of polymers in museums starts with optimal storage and display conditions. There is always a compromise between the ideal and the practical in this area. Because any level of light eventually causes damage to polymers, an ideal condition would be to keep the artifacts permanently in the dark. However, this negates their value as examples of their technology and their usefulness in educating the museum visitor – few people would visit a museum that had no artifacts on display. Research on display conditions has therefore centered around measuring the amount of damage caused by different light levels, temperatures, humidities, pollution levels, and mechanical stresses that are normally encountered in museums. Unfortunately, even a moderate set of suggested conditions, which allows adequate display with minimal damage, is often outside the budget of many museums. Further research in this area may be directed toward more economical methods of safe display.

Similarly, ideal storage conditions may be too expensive for many museums. The problem of compromising between accessibility to visitors and the health of the artifacts is minimal, as objects in storage are not usually expected to be viewed. However, if the collection is to be maintained, studied and periodically surveyed for any damage, it must still be accessible to the museum professionals. If an object needs to be accessed once a month, but is in a freezer and requires one week of slow warming for access and one week of slow cooling to be returned to the freezer, that object will be unlikely to be returned to the freezer – even the most dedicated professional will lapse if a storage regimen is inconsistent with access requirements.

Therefore, the benefits of various storage conditions must be quantified, both to balance against the cost of the storage and against accessibility needs. Tumosa, et al., investigated the different polymers found in photographic film and their properties in cold storage to help make such an assessment for photographic collections. The materials in such collections are generally more consistent than in collections of other artifacts, such as plastics and artwork, and some very specific predictions can be made. In some cases the prediction is complicated by the multi-layered nature of photographic materials; the materials must be looked at as a composite. Based on the work of Tumosa, et al., a curator can set a reasonable storage regimen, taking planned accessibility into account, and know the effects of the storage, the out-of-storage time and the changing conditions.

Testing New Materials

The use of modern polymers to preserve museum objects is perhaps the most interesting aspect of polymers in museums. For any need in preserving an object, there is likely to be a polymer that exists or can be designed to meet that need. Modern polymers, particularly acrylics, are currently used as protective coatings, adhesives and in display cases. Polyesters, particularly in film form, are effective protective barriers, and polyolefins are used for supports and storage containers. By using the right polymer, a museum professional can reduce or exclude moisture, oxygen, light and pollution; in many cases this can be done more economically than with other materials.

If a polymer is to be adhered to an object, as with a varnish, adhesive, consolidant or fill, its ideal properties are:

- long-term stability (on the order of decades)
- no irreversible interaction with the object
- ease of removal of the polymer from the object

The polymer should not change in appearance, nor change chemically with age. It should not react chemically with the object or have components (such as additives or by-products) that will migrate over time into the object. It should be removable at any time with no effect on the object. This last is very hard to accomplish – any solvent that will dissolve the polymer is likely to somewhat affect the object. Combining these necessary characteristics with desired characteristics -- such as a certain refractive index, clarity, hardness, or flexibility -- can make it impossible to find a commercial product that is suitable, and compromises are often made. However, as new synthetic routes and new polymers are developed, research on their potential for museum use should continue; ideal polymers may come from unlikely sources that should not be overlooked.

One aspect of the requirement for ease of removal is that the material must be easily identified as non-original. This is a great advantage of modern polymers when used on older objects: a repair to a Renaissance painting, when done with acrylics, is easily identified chemically. Similar care can be taken with polymers used on more recent works to incorporate materials that a scientist can identify as anachronistic.

Polymers that are used near or in contact with objects have fewer requirements; they still must not contain or produce components that will contaminate the objects, but long-term stability is otherwise a matter of economics rather than preservation. If a storage container is ideal otherwise, but must be replaced every ten years, it might still be a better choice than one that lasts longer but has other drawbacks.

Predicting the long-term properties of these polymers generally requires artificial aging. Some industrial testing can be used as a basis for choosing polymers, but that testing often is not concerned with the effects of the polymer on other materials. For example, a polyethylene film might be considered to be structurally stable for decades, according to the industrial testing, but if it contains certain antioxidants that can migrate into objects and cause staining over time, it can not be used for long-term storage. Once again, testing polymers that were used in the past for preservation may help develop materials that have all the ideal characteristics.

Polymers in museums, as objects that need preservation and study, and as materials for the preservation of all museum objects, will continue to benefit from advances in the general field of polymer science. As better techniques and new materials become available, they will find a place in museums as well as in industry. Further research on polymers in museums may sometimes benefit other fields as well as continue to improve our ability to preserve the objects that reflect our culture and history.

Chapter 11

The Physical Properties of Photographic Film Polymers Subjected to Cold Storage Environments

Charles S. Tumosa[1], Marion F. Mecklenburg[1], and
Mark H. McCormick-Goodhart[2]

[1]Smithsonian Center for Materials Research and Education,
Smithsonian Institution, Washington, DC 20560–0534
[2]Old Town Editions, Alexandria, VA 22314

The photographic record of the twentieth century is rapidly being lost. The gelatin of the image emulsions is relatively stable, but the silver salts and dyes that form the image, and the principal photographic film bases, cellulose nitrate and cellulose triacetate, have been found to be chemically unstable within the time frame of historical significance. In an effort to save these photographic records, several strategies to enhance their useful life have been developed. These involve low relative humidity (RH) and/or low temperatures. The ability of the polymeric film bases to sustain these lower temperatures has been examined. The data indicate that these materials can safely tolerate storage at low temperatures, easily to -20 C, and that cycling of the film bases within the range +25 C to -25 C has no adverse effect on the mechanical stability of the film bases.

The long term preservation of photographic materials must of necessity consider the adverse effects of light, relative humidity (RH) and temperature. Storing the film in the dark is almost axiomatic but the optimal values or ranges of RH and temperature have been the subject of much discussion. Previous research has described the benefits and tradeoffs between RH and temperature (1). From the chemical point of view (Arrhenius considerations), it is clear that any long term storage of photographic materials would benefit more from strategies involving low temperature than low RH. However, it was necessary to demonstrate that there are no structural hazards involved in the low temperature approaches. A number of questions require answers. What are the dimensional changes and thermal coefficient mismatches associated with lowering the temperature? Are significant stresses developed and do these stresses represent a potential for damage to the film? Finally, how often can the films be cycled through temperature changes in order to provide user access?

Dimensional Response to Thermal Changes

Studies of the mechanical properties of the photographic film bases, cellulose triacetate (recent origin) and nitrocellulose (50 years old) were conducted. For stability during storage it is required that any temperature related dimensional changes of the cellulose nitrate and cellulose triacetate polymers be such that there are no severe stresses induced in either the image or film base layers. The basic criterion for damage is that the stresses in any of the photograph materials should not exceed the yield point, the point beyond which permanent deformation occurs. For a large class of organic polymers the yield point has been found to be at a minimum strain value of 0.004. This was also found to be true for both the image emulsions and the different film bases. Figure 1 illustrates the full stress strain plot for 50-year-old nitrocellulose film stock base. The 0.004 yield point is a conservatively low value and allows one to set a specific limit to the minimum temperature change required to cause permanent deformation in a fully restrained specimen. It is important to note that the breaking stress and strain for the material far exceed the yield stress and strain used to calculate the allowable environmental limits and permit considerable leeway in handling films.

Figure 2 shows the dimensional changes of both cellulose triacetate and cellulose nitrate over the temperature range +40 C to -40 C and at a relative humidity of 50%. Similar curves have been developed at other relative humidities with the same general response. What is particularly interesting is that the dimensional response of both motion picture film bases is quite similar at the ambient temperature ranges expected in movie theaters, which is around 18 to 25 C. Perhaps this was the intent of the manufacturers since both films must fit the projector sprockets equally well. This physical match also extends down to about -20 C, the storage temperature of most modern single stage freezers.

Another interesting effect is the raising of the yield point on lowering the temperature. Further mechanical testing at low temperatures revealed that the yield point (strain) for nitrocellulose rose to as high as 0.011 at -20 C. The yield strains for this material over a wide range of temperatures are shown in Figure 3. All are above the 0.004 value used as the damage criterion. Damage by our definition is permanent deformation, not failure. Much greater strains are required to cause actual failure. Similar measurements have been made for gelatin, the principal component of the image layer, and have been reported elsewhere (1,2). Within the RH range of 30% to 60% gelatin also has yield points in excess of 0.004.

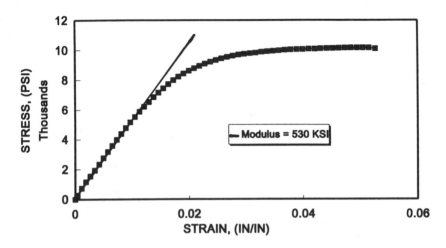

Figure 1. The full stress-strain plot for a 50 year old nitrocellulose film stock base.

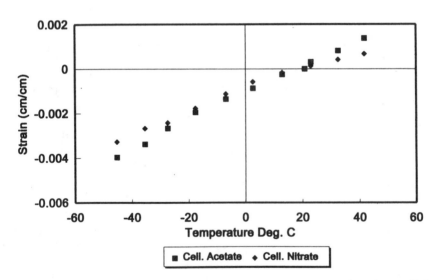

Figure 2. The dimensional response to temperature, as strain, of modern cellulose triacetate and 50 year old nitrocellulose film stock base. The RH is 50%.

Thermally Induced Stress and Strain Development in Restrained Films

Cooling a restrained photographic film will cause the development of both stresses and strains in the various layers of the film. That there are stresses developed in a material restrained from shrinking is intuitive. But strains are also developed. If a material of length L_i is allowed to freely shrink on cooling to a new length, L_f, the material is stress free. If, at the lower temperature, the material is stretched back to its original length, L_i, the material has both stress and strain. The values of stress and strain are identical to those developed if the material had been fully restrained and not allowed to shrink when cooled. It can be shown that the magnitude of the dimensional change of a material freely contracting during cooling is the same as the mechanical strain developed when the material is restrained during cooling. Further, the stresses developed during restrained cooling are identical to those developed when a material is stretched backed to its initial length after freely shrinking during a temperature reduction (4). In the case of motion picture film, the worst case assumption is that the film is fully restrained from shrinking when it is spooled on the reel. Full restraint is the worst possible condition though no structure is actually fully restrained. If the fully restrained material can safely tolerate thermal change under these conditions, then it can tolerate cooling under all other conditions of lesser constraint.

Assume that the film in question has a film base of cellulose triacetate. Figure 2 shows that when this material cools from 20 C to –20 C the magnitude of the free swelling strain is 0.002. If the material had been restrained, the magnitude of the mechanical strain would have been 0.002 also. This is only half the lowest permissible strain of 0.004 used as a criterion for damage. The film base can safely be lowered to –20 C. But is the gelatin emulsion layer also safe?

The thermal coefficient of expansion (the slope of the strain versus temperature plot of an unrestrained material) for the cellulose triacetate is approximately 0.00005/°C in the region of interest. For the gelatin image emulsion, the thermal coefficient of expansion is 0.000030/°C. If the emulsion were to be considered as an independent material alone and restrained, the same temperature drop, from 20 C to –20 C, would induce mechanical strains of only 0.0012, less than the film base. If the film is restrained, all layers are restrained and the response of each layer can be considered independently.

Cut film is normally stored unrestrained in sheets of different sizes. What is the response of the combined film base and the gelatin image layer during cooling? The mechanical strain developed in the different layers of a film when it is cooled unrestrained is easily approximated. The film base tends to contract (thermal coefficient of 0.00005/°C) at a greater rate than the gelatin emulsion layer (thermal coefficient of 0.00003/°C). The response of the composite is intermediate between the layers (if they were acting independently) with the actual response dependent on the relative thickness and stiffness of the different layers. For film, the base is so much thicker than the image layer that it essentially determines the response of the composite. Thus, the strains developed in the gelatin layer are approximately the temperature change times the difference in the two coefficients. In this case the thermal change is –40 C and the difference in coefficients is 0.00002/°C., so the emulsion layer develops a mechanical strain of 0.0008 in compression. This is well within the elastic range of the material. It would take a more sophisticated analysis to calculate the exact resulting strains but the answer would be less than computed above.

Thermal Cycling of Materials

The number of times that photographic materials can be cycled between ambient and cold storage temperatures also needs to be determined for an adequate storage strategy. If the strain values encountered during the cycling stay below the yield point, then no permanent deformation of the object takes place, regardless of the number of cycles it experiences. Since the changes in temperature produce strains that are within the elastic region (strain less than 0.004) then the concept of fatigue need not enter into consideration. Since it can be shown rigorously that mechanical and thermal cycling are identical (4), then any mechanical cycling to strains of 0.004 would reflect thermal cycling from 22 C to –40 C. Lower values of strain would correspond to a smaller temperature drop. Figures 4 and 5 show the mechanical cycling or unload compliance curves of nitrocellulose film and gelatin for selected points over a range of 5000 cycles. No plastic deformation was observed. Since the deformation exceeded the 0.004 value slightly, it is evident that this value is conservative for both materials. Similar results were obtained for the cycling of cellulose triacetate.

Discussion and Conclusions

Twentieth century photographic materials suffer from several different mechanisms of degradation, including the hydrolysis of cellulose acetate ("vinegar syndrome") and dark fading. In fact, they are often less stable than 19th century photographic objects. What these mechanisms have in common is that their rates are more sensitive to change in temperature than RH. Photographs at room temperature should be kept at a RH between 35% and 65% in order to avoid damaging stresses and strains produced by differential dimensional responses of the photographic layers to RH changes. The possibility of mold growth or going above the glass transition temperature, T_g, of the gelatin emulsion layer further reduces the preferred upper limit to 60% RH. Reducing the RH within the stated range can increase the chemical stability by a factor of 2 to 4, but this is not enough to ensure the survival of photographic collections. Lowering the temperature can produce orders of magnitude increases in chemical stability.

Truly long-term preservation efforts must consider factors that minimize the effects of several decay mechanisms and will invariably use strategies that lower storage or exhibit temperatures. The important question then is whether the photographic materials can survive the effects of cold temperature and of cycling into and out of storage environments for study or exhibition.

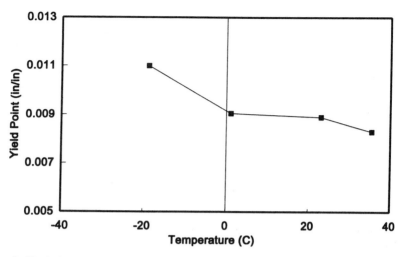

Figure 3. Variation in yield point with temperature for 50 year old nitrocellulose film stock base. The RH is 50%.

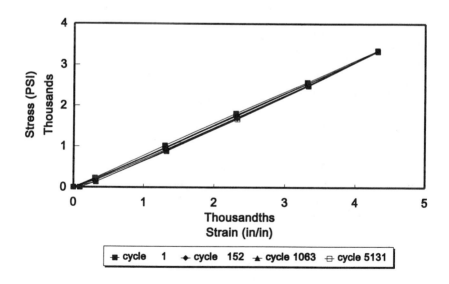

Figure 4. The unload compliance curves for nitrocellulose film stock base at various times during 5131 cycles to a strain level of 0.0044.

The data presented here indicate that the dimensional change induced by lowering the temperature to that of a single stage freezer, i.e. –20 to –30 C, is considerably less than that which would induce plastic deformation in either the polymeric base or the gelatin layers. The assumption that one or more layers in the photographic materials are fully restrained is conservative. In fact, all materials will be contracting together, with different coefficients of thermal expansion, so that relative strains will be even less than the data for individual materials indicate. Other objects such as glass plate negatives are a special case and are not being considered here although the same principles apply. The data also complement previous studies (2-4) that have shown that cycling within the ambient to –25 C range is benign.

The choice of –20 C for a storage temperature is an optimized value based on two independent principles. First, -20 C is a temperature easily attained in commercial freezers using single stage compressor systems. Second, if the collection is used to any extent during the year, then the effects of the colder storage must be time averaged with the effects of the ambient conditions of use. As the time of use increases, the degradation occurring during use becomes much greater than that during cold storage. The time-out-of-storage becomes the dominant consideration and the benefits of further improvement in stability by temperatures below –20 C during the time in cold storage become limited (1).

The data described above and similar data for other materials allow certain general conclusions to be drawn. As a practical matter photographic materials typically contain a gelatin image-bearing layer adhered to various substrates (e.g., cellulose nitrate or triacetate, glass, paper). The chemically and physically safe environment for these common objects can be summarized in a simple plot such as Figure 6. This figure shows the boundaries of RH and temperature, A, B, C, and D, which represent the limits of safe storage and exhibition. Point A is 35% RH, B is 60% RH, both at 25 C, while point C is 40% RH and D is 20% RH, both at -25 C.

The contour lines labeled 2, 4, 10, 30, 100, etc. denote the improvement in expected lifetimes compared to a room temperature environment of 21 C at 50% RH. The benefits of a preservation strategy that utilizes a lower RH, i.e., moving horizontally on the graph, has a maximum effect of increasing the expected life by a factor of 2 to 4. A strategy of lowering the temperature to sub-zero values, i.e., moving vertically on the chart, has a much more significant impact.

Hazards do exist with any storage or user environment that is beyond the glass transition of gelatin (as indicated in the graph). Procedures should be used to ensure that cold stored objects are returned to room temperature in such a way that the microclimate at the surface of the object does not condense water or otherwise exceed the recommended RH limits.

References

1. McCormick-Goodhart, M.H. J. Soc. Archiv. 1996, 17, 7.
2. Mecklenburg, M.F.; McCormick-Goodhart, M.H. and Tumosa, C.S. J. Am. Instit. Cons. 1994, 33, 153.
3. Mecklenburg, M.F.; Tumosa, C.S. and McCormick-Goodhart, M.H. Mat. Issues Art and Arch. IV 1995, 352, 285.

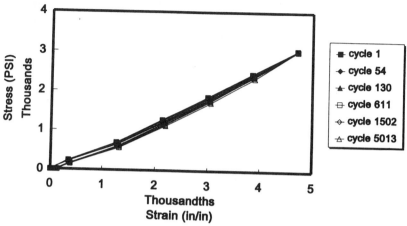

Figure 5. The unload compliance curves for gelatin at different times during cycling to a strain level of 0.0048. No yield is measurable.

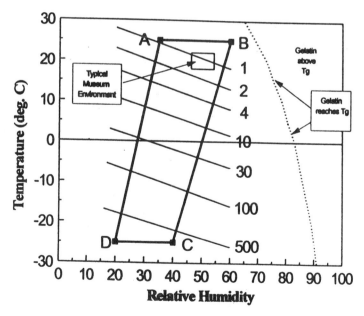

Figure 6. Range of allowable relative humidity and temperature for materials containing gelatin films.

4. Mecklenburg, M. F. And Tumosa, C. S., "The Relationship of Externally Applied Stresses to Environmentally Induced Stresses," Fiber Composites in Infrastructure, Proceedings of the First International Conference on Composites in Infrastructures, ICCI'96, H. Saadatmanesh and M. R. Ehsani, Eds. Tucson Arizona, (1996), 956-971.
5. McCormick-Goodhart, M.H. and Mecklenburg, M.F. IS&T's 46th Annual Conf. 1993, 277.

Chapter 12

Probing the Factors That Control Degradation in Museum Collections of Cellulose Acetate Artefacts

Jane Ballany[1], David Littlejohn[1,3], Richard A. Pethrick[1], and Anita Quye[2]

[1]Department of Pure and Applied Chemistry, University of Strathclyde, 295 Cathedral Street, Glasgow G1 1XL, United Kingdom
[2]National Museums of Scotland, Chambers Street, Edinburgh EH1 1JF, United Kingdom

Cellulose acetate artefacts in museum collections cover a period from the early 1900's to the present day. Conservators have observed that certain of these objects are showing signs of warping, crazing, cracking, discolouration and shrinkage accompanied by a strong smell of acetic acid. Previous studies on cellulose nitrate artefacts show a correlation existed between the residual sulfate from manufacture and subsequent susceptibility to degradation. A parallel study of the accelerated ageing of modern samples of cellulose acetate and also selected artefacts dating from the 1940's has been carried out. The tests involved exposure of the objects to temperatures of 35 °C, 50 °C and 70 °C and relative humidities of 12 %, 55 % and 75 % for extended periods of time. The samples were monitored for changes both in their visual appearance, mass and chemical composition. Chemical analysis was carried out using micro FT-IR spectrometry and ion chromatography. The changes in the molar mass distribution were studied using gel permeation chromatography. Naturally aged samples have also been studied to help validate the accelerated ageing studies.

Plastics have had an increasing influence on human activity since the early years of the 20th century and as such are becoming an increasingly important part of

[3]Corresponding author.

museum collections. However, a large number of artefacts in these collections are showing various signs of degradation. The synthetic plastics, including cellulose acetate, from the first half of the century are especially sensitive to these problems and therefore, investigation into the causes of degradation is essential.

Cellulose acetate was first discovered in 1865 by Schutzenberger when he heated cotton wool and acetic acid to 140 °C in a sealed tube (1). Manufacturing processes in the 1920's - 1940's were slightly more sophisticated, using lower temperatures and adding sulfuric acid as a catalyst (2). Acetic anhydride also replaced acetic acid as the acetylating agent, owing to its better efficiency. Furthermore, the use of pressurised vessels in recent years has reduced the manufacturing time from several days (3) to hours (4). Although sulfuric acid is a useful catalyst, incomplete washing of the product can result in some of the acid being trapped within the plastic, which may be a source of degradation, as occurs in certain cellulose nitrate artefacts (5). It was shown that with cellulose nitrate, chain scission was the main degradation pathway, initiated by hydrolysis of residual sulfate by water to sulfuric acid.

Conservators have observed that certain cellulose acetate artefacts are showing signs of warping, crazing and shrinkage accompanied by a strong smell of acetic acid and a sticky surface. Discoloration is also a major problem, especially with doll's. Previous studies on cellulose acetate films (6-8) have shown that degradation may be caused by deacetylation, hydrolysis and/or plasticiser loss.

Deacetylation produces acetic acid vapours, known as the vinegar syndrome (6,7,9), which cause further problems. The acid vapours released to the atmosphere can cause the 'spread' of the 'disease', commonly known as "Doll's Disease" (10,11), as the acid will catalyse the deacetylation of other undegraded cellulose acetate artefacts.

The trapped acid can also catalyse hydrolysis of C-O bonds in the polymer backbone causing a reduction in the polymer chain length, known as chain scission, with subsequent weakening of the polymer (12). This chain scissioning can be triggered by ultra-violet or visible light (13,14) then catalysed by trapped acetic or sulfuric acid. The cellulose polymer backbone is susceptible to chain scission as the lactose ring can undergo acid catalysed opening, leaving the polymer susceptible to bond dissociation resulting in a build up of oxalate residues on the artefact.

20 – 40 % (w/w) plasticiser is added to cellulose acetate (11) to allow it to be used as a mouldable plastic. Hence, plasticiser loss can be a major factor in the degradation of cellulose acetate artefacts, even although the polymer is not directly changed. Plasticiser loss has two major affects: a) the surface becomes sticky and unpleasant to touch; b) the plastic becomes brittle and more susceptible to physical damage. It is thought that plasticiser loss occurs after degradation of the polymer, because it is believed to become insoluble in the deacetylated polymer matrix and migrates to the surface (10). A separate study of a Gabo sculpture (13) observed that

a vinegar odour was detected five years before an oily substance, i.e. plasticiser, appeared on the surface.

A study of the accelerated ageing of cellulose acetate samples has been used to assess the dominant process(es) involved in cellulose acetate degradation and understand the factors which affect the rate of degradation. The samples selected for study included artefacts from the period 1940 - 70 and modern test pieces. The procedure has also been validated by studying naturally aged samples from various museums. The main aims of the study were to (a) determine the residual levels of sulfate and other ions on the surface of an artefact, (b) understand more fully the pathways involved in the degradation of cellulose acetate and (c) investigate the storage and/or display conditions most likely to cause degradation.

EXPERIMENTAL

Samples

Samples for the accelerated ageing study were taken from artefacts covering a range of ages and degrees of degradation. The samples used were:
(1) A doll from the 1940's - (a) the body was in excellent condition showing no visible signs of degradation, (b) the leg was badly discoloured from pink to orange with large areas of crazing and cracks; the surface was also blistering.
(2) A tortoiseshell effect comb manufactured in 1946; in good condition with only a few small areas of crazing.
(3) A tortoiseshell effect comb manufactured in 1967; in excellent condition showing no visible signs of degradation.
(4) A tortoiseshell effect hair slide manufactured in 1967; in excellent condition.
(5) A transparent test piece of cellulose acetate manufactured in 1996 by Courtaulds Chemicals; used as the 'control sample'.

Samples were cut into pieces approximately 4 cm x 5 cm (thickness ranged from 2 – 5 mm), weighed and suspended from the lid of a small dessicator (volume about 950 mL), using polystyrene strips and small pieces of 'blu-tac'. A saturated solution of a particular salt was placed in the well of the dessicator to control relative humidity (RH), as illustrated in figure 1. The dessicators were exposed to temperatures of 35 °C, 50 °C or 70 °C in ovens and the relative humidities in the dessicators were controlled at 12 % (lithium chloride), 55 % (magnesium nitrate) or 75 % (sodium chloride).

A piece of each artefact was placed in each dessicator resulting in six samples per dessicator. This meant that samples of each artefact were exposed to all temperatures and relative humidities. The experiments at 70 °C were stopped after 62 days exposure, due to the extreme degradation of the samples. The experiments at 50 °C and 35 °C are still being conducted (in excess of 500 days) as of July 1999.

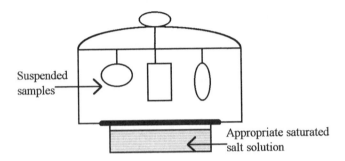

Figure 1: Illustration of accelerated ageing apparatus

The naturally aged samples studied were collected from various US museums and were as follows:

(1) A lamp manufactured by Artemide; fine cracks were beginning to appear at edges.
(2) An Inguis Box designed by René Jules Lalique in 1924; showing blistering, cracks and fine white crystals, date of manufacture uncertain.
(3) A Moholy-Nagy "light modulator painting" 1938; smells strongly of acetic acid, badly warped and has areas which exhibit cracks and crazing.
(4) A storage folder used to house fabric; beginning to warp and with a slight smell of acetic acid.
(5) Various combs from a collection of mixed combs; signs of warping, cracking and discolouration.
(6) A Duchamp painted acetate image of a large glass sculpture; showing slight warping.

Analyses

Solubility tests

The generic term of 'cellulose acetate' has been used as the commercial name for four cellulose esters: cellulose diacetate, cellulose triacetate, cellulose acetate butyrate and cellulose acetate propionate. These four polymers all emit acetic acid on degradation and produce similar, though distinguishable FT-IR spectra.

However, the solubility of each is very distinct: cellulose diacetate is readily soluble in acetone, cellulose triacetate in dichloromethane and cellulose acetate butyrate does not dissolve in either solvent. Cellulose acetate propionate was only produced as a speciality polymer and was not commonly used. Therefore, solubility was used as a means of identifying the exact form of 'cellulose acetate' found in each sample used for accelerated ageing. Unfortunately these tests were not possible on any of the naturally aged samples as these are all museum artefacts and therefore, pieces could not be removed for solubility testing.

Tests were conducted by placing approximately 0.5 g of each sample in 5 mL acetone or 5 mL dichloromethane. The samples were then left for 24 h and the solubility of each sample was noted.

Micro FT-IR spectroscopy (FT-IR)

FT-IR was used to confirm the identity of the base polymer of the plastic and the plasticiser before testing began and also to monitor changes in the magnitude of the ratio of the O-H or C=O peaks compared with C-H peaks at 3450, 1750 and 1370 cm^{-1}, respectively, as ageing progressed. An MCTA Nicplan microscope was coupled to a Nicolet 510P spectrometer. A 10X magnification objective lens was used to locate regions for analysis and a Cassegrainian Reflechromat objective lens (SpectraTech) with 15X magnification was used for collection of spectra. Each spectrum was collected with 128 scans over the wavenumber range 4000 - 650 cm^{-1}, with a resolution of 8 cm^{-1}.

The plastic was identified using the transmission mode. A microscopic piece of sample was removed from the artefact using a scalpel and crushed between two diamond windows. The spectrum was recorded through only one diamond window which supported the sample. Both planar and faceted diamonds have been used to allow comparison to be drawn between the two techniques.

The plasticiser was identified using the reflectance mode. A cotton wool swab (The Boots Company plc, Nottingham, England) moistened with dichloromethane was wiped several times over a 2 cm^2 area of the artefact surface. The plasticiser was extracted from the swab into 2 mL of dichloromethane, which was then evaporated to a small volume. This was transferred onto a gold mirror and evaporated to dryness and an IR spectrum recorded. A background spectrum was collected from a blank cotton wool swab and subtracted from all other spectra recorded.

Ion chromatography (IC)

IC was used to investigate the presence of acetate, formate, chloride, nitrate, sulfate and oxalate in the plastic. An increase in acetate levels indicates that deacetylation is occurring, oxalate shows chain scission is involved in ageing and formate is produced by oxidative degradation. Chloride and sulfate are anions from residual chemicals from the manufacturing process, as chloride salts are often used to stop acetylation, while sulfuric acid is the most common catalyst used. Higher nitrate concentrations may indicate the use of a cellulose acetate - cellulose nitrate mixture to give desired colour effects.

A Dionex 4000i chromatograph with AG4A guard column and AS4A separator column and a 6 mM sodium tetraborate (borax) eluant were used to separate acetate, formate and chloride. A Dionex DX-100 chromatograph with AG5 guard column and AS5 separator column and sodium carbonate : sodium bicarbonate (2.2 mM : 2.8 mM) eluant were used to separate chloride, nitrate, sulfate and oxalate. The naturally aged museum samples were analysed on a Dionex 500 Chromatograph with AG4A-SC guard column and AS4A-SC separator column and a borax gradient elution method. The gradient method consisted of an initial 2 minute rinse with distilled water followed by 11.5 minutes at 4.2 mM borax. The eluant strength was then increased linearly to 22.75 mM borax over 7 minutes and held for a further 7 minutes before a final 2 minute rinse with 4.2 mM borax. This allowed all six anions to be determined in one analysis. Suppressed conductivity detection was used in all cases.

Samples were prepared either destructively or non-destructively. Destructive sampling involved the removal of approximately 50 mg of material, which was weighed accurately and soaked in 5 mL distilled water at room temperature for 24 h. The aqueous extract was then analysed by ion chromatography. Detection limits were calculated as acetate 0.34 $\mu g \, g^{-1}$, formate 0.27 $\mu g \, g^{-1}$, chloride 0.17 $\mu g \, g^{-1}$, nitrate 0.39 $\mu g \, g^{-1}$, sulphate 0.22 $\mu g \, g^{-1}$ and oxalate 0.12 $\mu g \, g^{-1}$.

Non-destructive sampling involved a cotton wool swab (The Boots Company plc, Nottingham, England), moistened with distilled water, being wiped twenty times over a 2 cm^2 area of the artefact. The cotton wool swab was soaked in 5 mL distilled water for 10 minutes; the extract was then analysed by ion chromatography. Detection limits were calculated as acetate 0.74 $\mu g \, cm^{-2}$, formate 0.41 $\mu g \, cm^{-2}$, chloride 0.49 $\mu g \, cm^{-2}$, nitrate 0.99 $\mu g \, cm^{-2}$, sulphate 0.66 $\mu g \, cm^{-2}$ and oxalate 0.94 $\mu g \, cm^{-2}$.

In both procedures the levels of anions present were quantified, in concentrations of $\mu g \, g^{-1}$ for the destructive method and $\mu g \, cm^{-2}$ for the non-destructive method. Non-destructive sampling was used from all naturally aged samples.

Gel Permeation Chromatography (GPC)

GPC was used to monitor changes in the molar mass distribution of the polymer in an artefact during accelerated ageing. The apparatus used was a Pye Unicam PU 4003 pump in conjunction with a Pye Unicam PU 4003 controller. A tetrahydrofuran (HPLC grade) (THF) eluant was passed at a flow rate of 1 mL min^{-1} through two GPC columns in series, packed with 5 μm PL gel particles of 10^4 Å and

500 Å pore size. Separated components were monitored using a Pye Unicam PU 4026 refractive index detector.

Samples were prepared by dissolving approximately 20 mg of plastic in 20 mL THF with 0.5 % (V/V) toluene as an internal standard. The samples were left for at least 24 h to ensure complete dissolution.

RESULTS AND DISCUSSION

Solubility tests

The solubility tests (table I) showed that all of the samples used were readily soluble in acetone and therefore, can be positively identified as cellulose diacetate (referred to as cellulose acetate throughout). A small amount of insoluble white powder, was left in the acetone solution for both the doll's body and leg, indicating that a small amount of the original material was not cellulose acetate; this is most likely to be TiO_2 filler used to produce an opaque plastic.

Table I: Solubility Test Results

Sample	Solubility		'Cellulose Acetate' Type
	Acetone	Dichloromethane	
Doll's body	✓	✗	Diacetate
Doll's Leg	✓	✗	Diacetate
1946 comb	✓	✗	Diacetate
1967 comb	✓	✗	Diacetate
1967 piece	✓	✗	Diacetate
1996 ref. CA	✓	✗	Diacetate

Visual

Changes in colour, size and pliability were observed during accelerated ageing. All of the samples visibly shrank after long term exposure (62 days) at 70 °C. The samples became very brittle and disintegrated when touched. Degradation, especially embrittlement, was fastest at a high relative humidity, indicating that a high moisture content in the air accelerates processes which lead to plasticiser migration.

The samples aged at 50 °C (594 days) showed similar signs of degradation to those at 70 °C, although the transformation was slower, allowing individual stages to be identified. A strong smell of acetic acid was observed early in the ageing process, around 100 days, followed by slight discoloration and a sticky surface after approximately 200 days. The samples then progressed to become severely discoloured and brittle. At 35 °C, only the samples at 75 %RH showed signs of degradation after 500 days, with the surface becoming sticky and discoloured, only a very slight odour of acetic acid and plasticiser was detected. This suggests that degradation is far slower than at the higher temperatures.

Weight Loss

The percentage weight losses for all accelerated aged samples are shown in table II. A significant weight loss was observed in all samples at 70 °C and 50 °C, however those at 35 °C showed little change in weight. Figures 2 and 3 show the percentage weight loss during accelerated ageing for the 1996 reference sample and the doll's leg, respectively.

Table II: Percentage weight loss of samples after ageing

Sample	Weight loss / %								
	70 °C (62)			50 °C (594)			35 °C (498)		
	75 %RH	55 %RH	12 %RH	75 %RH	55 %RH	12 %RH	75 %RH	55 %RH	12 %RH
Doll's body	52.9	21.5	22.6	8.6	1.7	2.7	2.0	0.5	1.4
Doll's Leg	55.1	40.2	42.0	10.7	-4.9	1.5	6.4	-1.8	3.6
1946 comb	35.1	15.4	0.7	28.7	11.9	8.1	-0.9	2.8	2.2
1967 comb	28.8	22.8	14.3	26.5	7.5	3.0	0.04	-0.4	1.6
1967 piece	16.1	9.3	6.9	7.0	2.0	2.8	-1.5	-1.6	1.6
1996 ref. CA	16.3	13.9	8.3	13.0	17.5	5.5	-1.0	-0.5	1.7

( ) – number of days exposure

The weight loss was due to two processes, loss of acetate and loss of plasticiser to the atmosphere. The modern cellulose acetate material showed a weight loss of 16.3 % over 62 days at 70 °C and 75 %RH (initial weight 12.137 g). The older samples generally showed a greater weight loss than the modern cellulose acetate, which was most likely caused by a faster rate of degradation.

Over 594 days at 50 °C, all samples exhibited a weight loss of between 7.0 and 28.7 %, the only exception being the doll's leg which gained weight. This weight gain was due to absorption of some of the magnesium nitrate salt solution which was

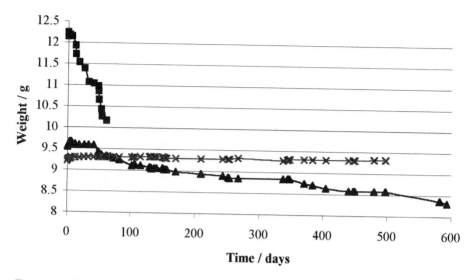

Figure 2: Change in weight of 1996 reference cellulose acetate during ageing at 75 %RH and 70 °C (■), 50 °C (▲) and 35 °C (×).

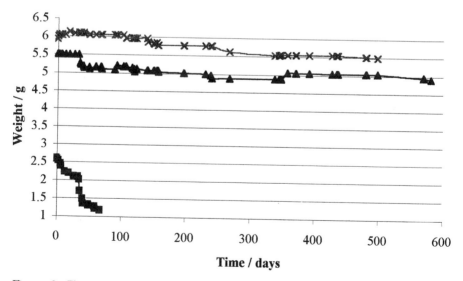

Figure 3: Change in weight of doll's leg during ageing at 75 %RH and 70 °C (■), 50 °C (▲) and 35 °C (×).

confirmed by a sudden and large increase in the nitrate concentrations detected by ion chromatography.

The results for ageing at 35 °C show that only the doll's leg at 75 %RH exhibited a significant weight loss (6.4 %).

Micro FT-IR spectroscopy (FT-IR)

FT-IR spectra identified all of the samples as cellulose acetate with a phthalate plasticiser. The FT-IR spectrum of the extract taken from the surface of the 1946 comb is shown in figure 4, along with the spectra for dioctyl- and dimethyl-phthalate. These spectra appear to indicate that the extract is predominately dioctyl-phthalate. However, plasticisers do not stay blended easily with cellulose diacetate and only those with relatively low molecular weight can be used. Therefore, it is unlikely that dioctyl-phthalate was the original plasticiser.

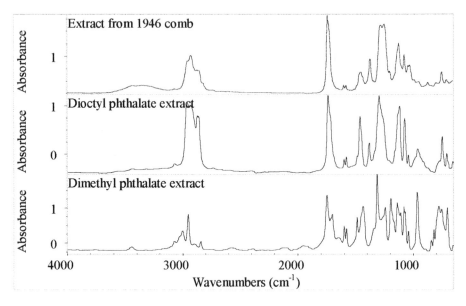

Figure 4: FT-IR spectra of plasticiser extracts

As ageing of all the cellulose acetate samples progressed there was an increase in the O-H peak intensity at 3450 cm^{-1} accompanied by a decrease in the C=O peak size at 1750 cm^{-1} and a shift in the peak position from 1750 to 1720 cm^{-1}. This is illustrated in figure 5 for the doll's leg during ageing at 50 °C and 75 %RH. These observations indicated deacetylation of the cellulose acetate to cellulose. Ratios of the magnitude of both O-H and C=O peaks compared with the C-H peak were calculated from the absorbance spectra and are shown in table III, along with peak position and full width at half height (FWHH) for the O-H and C=O peaks.

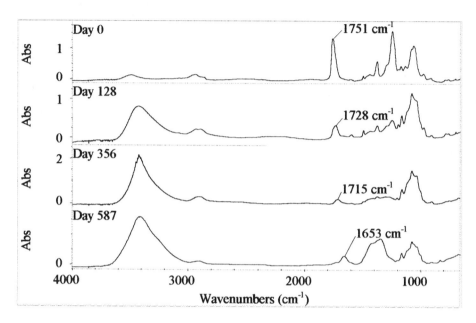

Figure 5: FT-IR spectra tracing the progression of degradation and subsequent shift in carbonyl band in the spectra of the doll's leg sample after ageing at 50 °C and 75 %RH: (a) – 0 days, (b) - 128 days, (c) - 356 days, (d) - 587 days.

The results, in table III, show that many differences can be seen between the spectra of degraded and undegraded samples. There is a large increase in the O-H peak along with a decrease in the C=O peak (both relative to the C-H peak) as degradation progresses. There is also an increase in the width of both peaks as degradation occurs. The ratios of the O-H to C=O peaks in the spectra of degraded samples resemble that of cellulose more closely than cellulose acetate, indicating that the polymer is returning to the original cellulose. There is also a significant shift in the position of the C=O peak from 1754 cm^{-1} to 1720 cm^{-1} as degradation occurs, which also indicates that the cellulose acetate is returning to the original cellulose.

Table III: FT-IR Peak Ratios and Positions

Sample Name	Ratios *		Position cm⁻¹		FWHH⁺ cm⁻¹	
	O-H	C=O	O-H	C=O	O-H	C=O
Reference cellulose	3.35	-	3397.0	-	257.4	-
Reference cellulose acetate	0.05	2.48	3471.2	1754.9	114.4	64.35
Cellulose acetate square (not degraded)	0.65	1.72	3465.5	1753.0	271.7	57.2
Cellulose acetate square (degraded)	3.12	0.76	3409.5	1722.1	343.2	78.7
Storage folder (slightly degraded)	0.47	1.70	3473.2	1745.3	271.7	57.2
Moholy-Nagy (degraded)	4.06	0.03	3394.1	1720.2	400.4	71.5
Doll's leg 70 °C 12 %RH day 62	0.88	1.54	3469.3	1743.3	300.3	57.2
Doll's leg 70 °C 75 %RH day 62	2.47	0.91	3423.0	1720.2	357.5	71.5

*Ratio to C-H at 1370 cm⁻¹.

+ FWHH - full width at half height.

Ion chromatography (IC)

Analysis by ion chromatography, using the destructive sampling method, showed an increase in water-extractable acetate levels after 102 days, however there was no significant increase in water-extractable oxalate concentrations, except for the doll's leg (table IV A & B). This indicates that deacetylation occurs, rather than chain scission, in the initial degradation of cellulose acetate. The presence of oxalate would have suggested a greater importance of chain scission, as is the case with cellulose nitrate (5). However as ageing progressed further an increase in water-extractable oxalate concentrations occurred in all samples; as well as a further increase in water-extractable acetate levels. This suggests that chain scission becomes more important as degradation progresses.

The results from the non-destructive sampling method show an increase in water-extractable oxalate levels at the surface as well as an increase in water-extractable acetate levels after both 130 and 582 days (table V A & B). This indicates that chain scission may occur firstly as a surface reaction then advance into the bulk of the sample.

The levels of water-extractable sulfate and chloride in both sampling methods increased as ageing occurred, however, neither showed correlation with the degree of

Table IV A: IC Analyses of Samples from the Accelerated Ageing Study at 50 °C and 75 %RH - Destructive Sampling

Sample	Concentration[+] / $\mu g\ g^{-1}$								
	Acetate			Formate			Chloride		
	Day 0	Day 102	Day 582	Day 0	Day 102	Day 582	Day 0	Day 102	Day 582
Average blank	0.07	0.1	0.1	1.3	0.03	0.1	0.1	0.1	0.09
Doll's body	1.2	3316	3246	2.1	277.0	ND	9.3	21739	474
Doll's leg	21.9	2865	2102	ND	ND	ND	12.8	8383	3406
1946 comb	49.0	46.0	1736	ND	ND	2.1	9.2	26.4	411
1967 comb	6.8	35.3	2024	ND	ND	ND	4.5	13.9	286
1967 piece	24.4	16.1	3300	ND	ND	ND	8.1	11.8	34.8
1996 ref. CA	10.6	356.5	3331	ND	4.7	ND	8.1	190.9	72.6

+ SD values, obtained from the calibration error (n = 4), were < 5 % for all analyses. ND – not detected.

All results are blank corrected.

Table IV B: IC Analyses of Samples from the Accelerated Ageing Study at 50 °C and 75 %RH - Destructive Sampling

Sample	Concentration[+] / $\mu g\ g^{-1}$								
	Nitrate			Sulfate			Oxalate		
	Day 0	Day 102	Day 582	Day 0	Day 102	Day 582	Day 0	Day 102	Day 582
Average blank	0.3	0.4	0.06	0.2	0.04	0.03	0.04	0.08	0.04
Doll's body	6.1	34.6	10.4	3.9	5.5	10.3	ND	12.1	24.3
Doll's leg	1.4	10.3	2.2	32.4	224.1	69.5	1.9	239.0	154
1946 comb	6.0	5.0	2.9	2.4	1.1	41.8	ND	ND	13.2
1967 comb	2.3	13.1	0.8	1.4	ND	37.2	ND	ND	5.7
1967 piece	13.7	7.9	4.6	1.8	ND	11.2	ND	ND	2.9
1996 ref. CA	21.9	5.2	6.2	1.7	1.4	15.3	ND	1.6	9.7

+ SD values, obtained from the calibration error (n = 4), were < 5 % for all analyses. ND – not detected.

All results are blank corrected.

Table V A: IC Analyses of Samples from the Accelerated Ageing Study at 50 °C and 75 %RH – Non-Destructive Sampling

Sample	Concentration[+] / $\mu g\ cm^{-2}$								
	Acetate			Formate			Chloride		
	Day 0	Day 130	Day 582	Day 0	Day 130	Day 582	Day 0	Day 130	Day 582
Average blank	2.3	0.2	16.2	0.7	1.6	13.6	1.6	1.7	3.5
Doll's body	1.1	17.2	93.3	ND	ND	4.9	0.7	8.9	81.5
Doll's leg	1.5	16.3	71.3	ND	ND	2.9	1.8	28.2	132
1946 comb	2.4	ND	70.8	0.5	0.1	2.1	0.9	1.2	33.5
1967 comb	6.9	ND	56.4	2.1	ND	ND	2.5	1.8	7.5
1967 piece	0.9	7.0	19.0	0.1	0.4	2.1	ND	2.1	2.9
1996 ref. CA	ND	13.1	21.6	0.1	0.6	ND	ND	8.0	5.0

+ SD values, obtained from the calibration error (n = 4), were < 5 % for all analyses. ND – not detected.

All results are blank corrected.

Table V B: IC Analyses of Samples from the Accelerated Ageing Study at 50 °C and 75 %RH – Non-Destructive Sampling

Sample	Concentration$^+$ / $\mu g\ cm^{-2}$											
	Nitrate			Sulfate				Oxalate			
	Day 0	Day 130	Day 582	Day 0	Day 130	Day 582	Day 0	Day 130	Day 582			
Average blank	1.8	1.6	3.4	1.5	1.9	1.6	0.3	0.1	0.9			
Doll's body	0.1	ND	0.3	0.7	0.3	2.6	ND	1.8	4.9			
Doll's leg	2.6	0.2	ND	1.5	0.7	3.1	0.6	10.6	3.4			
1946 comb	0.4	17.3	ND	0.8	ND	2.0	ND	0.2	0.5			
1967 comb	1.3	8.7	0.5	1.6	0.3	3.8	ND	ND	0.2			
1967 piece	0.9	0.4	ND	0.2	0.4	0.5	ND	0.4	0.3			
1996 ref. CA	ND	1.3	0.6	ND	0.9	0.6	ND	0.8	1.4			

+ SD values, obtained from the calibration error (n = 4), were < 5 % for all analyses. ND – not detected.

All results are blank corrected.

degradation, suggesting that the residues of these ions from manufacture have little or no affect on degradation of cellulose acetate. Again this differs from what was concluded for cellulose nitrate (5) when a link was indicated between sulfate levels and the extent of degradation. In this case, sulfuric acid, hydrolysed from trapped sulfate, is the catalyst for chain scission.

Water-extractable formate levels were low for all samples, suggesting minimal occurrence of oxidative degradation. Concentrations of water-extractable nitrate varied throughout the samples and also as ageing progressed, indicating that the most likely source would be contamination

Table VI: Comparison of IC Analyses for Naturally Aged and Accelerated Aged Samples – Non-Destructive Sampling

Sample	Concentration$^+$ / $\mu g\ cm^{-2}$					
	Acetate	Formate	Chloride	Nitrate	Sulfate	Oxalate
Average blank	2.3	0.7	1.6	1.8	1.5	0.3
Doll's body *	93.3	4.9	81.5	0.3	2.6	4.9
Doll's leg *	71.3	2.9	132.1	ND	3.1	3.4
1946 comb *	70.8	2.1	33.5	ND	2.0	0.5
1967 comb *	56.4	ND	7.5	0.5	3.8	0.2
1967 piece *	19.0	2.1	2.9	ND	0.5	0.3
1996 ref. CA *	21.6	ND	5.0	0.6	0.6	1.4
Average blank	16.5	15.9	14.2	11.9	4.4	7.7
Artemide Lamp	228.1	ND	ND	ND	6.4	1.0
Lalique Box	143.0	47.7	14.8	2.3	6.8	2.8
Moholy-Nagy	314.8	13.5	ND	ND	ND	1.7
Storage folder	119.3	1.2	31.0	3.5	1.5	0.4
Average Comb	74.7	ND	ND	17.7	ND	1.7
Duchamp	243.3	38.0	40.8	ND	7.2	2.7

*Accelerated ageing at 50 °C and 75 %RH for 582 days.

+ SD values, obtained from the calibration error (n = 4), were < 5 % for all analyses.

ND – not detected.

All results are blank corrected.

Results from naturally aged samples show similar trends to those of the accelerated ageing samples (see table VI). However, the levels of acetate found on the naturally aged samples were always higher than those for the accelerated aged samples. This is a consequence of the less severe ageing conditions as the higher temperatures used for accelerated ageing will volatilise more of the acetic acid than would be the case for an artefact at room temperature. The oxalate levels are similar showing a similar level of surface chain scission in all samples. The other anions have very similar levels between natural and accelerated aged samples indicating the concentrations of contaminants from manufacture are consistent throughout the samples. The higher levels of chloride in the samples from the doll are most likely due to the handling of this artefact during its lifetime.

It is important to comment on the differences in average blank found between the accelerated aged and naturally aged samples. This is due to the use of a different type of cotton wool swab as transportation was required after sampling of the naturally aged samples. The swabs used had wooden stems and were contained in a sterile plastic case (Greiner Labortachnik Ltd, UK) resulting in detection limits as follows acetate 3.83 μg cm^{-2}, formate 3.16 μg cm^{-2}, chloride 5.79 μg cm^{-2}, nitrate 2.77 μg cm^{-2}, sulphate 2.38 μg cm^{-2} and oxalate 1.22 μg cm^{-2}.

Gel Permeation Chromatography (GPC)

Results obtained by gel permeation chromatography indicated that no change in the molar mass distribution of the polymer was apparent, however, the poor solubility of many of the aged samples caused difficulty with this technique. The results of both an aged and unaged sample of 1967 comb and 1996 reference cellulose acetate are shown in figure 6.

The change in solubility as the samples aged does, however, collaborate the theory that deacetylation is the main process of degradation, as the solubility of cellulose acetate is well documented as being dependent on degree of substitution (15), becoming poorer in THF as the cellulose acetate becomes more polar (i.e. less acetylated). Also, from visual examination of both cellulose acetate and cellulose nitrate samples, it can be seen that crazing is more prevalent in cellulose nitrate. This is connected to chain scission, as crazing is generally caused by a loss in molecular weight. If chain scission is not occurring as the primary degradation mechanism in cellulose acetate as indicated by the IC results, then it can be expected that crazing would be less prevalent.

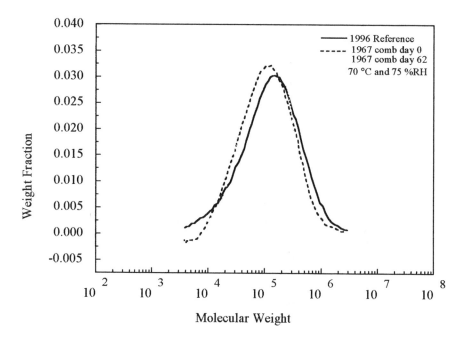

Figure 6: Molecular weight distributions

CONCLUSIONS AND FURTHER WORK

The IC results indicate that there is no correlation between residual sulfate or chloride and the extent of cellulose acetate degradation. This would be expected as the results from all of the analytical techniques used indicate that the predominant degradation process is deacetylation and this is not controlled by the hydrolysis of sulfate as is the case with the chain scission of cellulose nitrate. The increase in oxalate in older or more aggressively aged samples indicates that chain scission is a secondary process which only occurs later in cellulose acetate degradation. However, use of the non-destructive sampling method showed an increase in oxalate levels at the surface suggesting that chain scission could start as a surface reaction and, therefore, not be as important in 3D objects as deacetylation. The smell of acetic acid is detected before the surface of the artefact becomes sticky, with plasticiser, indicating that deacetylation occurs before plasticiser loss. It is probable that the loss of acetate changes the polymer matrix resulting in the plasticiser becoming insoluble and migrating to the surface. This is an important process in

164

degradation causing weight loss and loss of pliability, resulting in shrinkage and eventual destruction of the artefact.

Increases in temperature greatly increase the rate of degradation via deacetylation and plasticiser loss. At each temperature, weight loss increased with relative humidity indicating that a high moisture content in the atmosphere increases plasticiser migration and loss resulting in embrittlement of the plastic. Therefore, lower temperatures and relative humidities should be used when storing and displaying cellulose acetate. The storage and display areas also need to be well ventilated to allow acetic acid vapours to be removed; the use of scavengers could also be helpful.

Accelerated ageing studies will be continued at 35 °C. Headspace gas chromatography will also be used to determine if weight loss is due mainly to acetate or plasticiser loss by identifying and quantifying levels of both acetic acid and phthalate plasticiser in the atmosphere surrounding the samples. Also thermal analysis coupled to FT-IR will be carried out to help identify the degradation products as the sample is heated.

Overall, it can been seen from all the analysis that samples showing greatest visual changes also have a larger weight loss, increase in water extractable acetate concentrations and reduction in the IR spectrum carbonyl peak. These results show a link between acetate loss and visual degradation of the artefact as well as indicating that plasticiser loss occurs after the loss of acetate. However the lack of chain scission means that the samples can remain physically intact, even when shrinkage occurs and shows that with careful handling these artefacts can be saved for the future.

ACKNOWLEDGEMENTS

Jane Ballany thanks the EPSRC, Conservation Bureau of Historic Scotland and the National Museums of Scotland for their financial support throughout this project. Also thanks to Metropolitan Museum of Art (New York), Peabody Museum (Harvard), Philadelphia Museum of Art and Los Angeles County Museum of Art for supplying samples and the Smithsonian Institute (Washington DC) for the use of IR and IC equipment.

REFERENCES

1. Buttery, D.N. *Cellulose Plastics*; Cleaver Hume Press Ltd: London, 1947; pp. 87-112.
2. Stannett, V. *Cellulose Acetate Plastics;* Temple Press Ltd: London, 1950; pp. 22-41.

3. Barron, H. *Modern Plastics;* Chapman and Hall Ltd: London, 1945; pp. 314-338.

4. Personal correspondence, Courtaulds Manufacturing Scheme.

5. Stewart, R.; Littlejohn D.; Pethrick R.A.; Tennent N.H.; Quye A. In *From Marble to Chocolate – The Conservation of Modern Sculpture*; Heuman J., Ed.; Archetype Publications: London, 1995; pp. 93-97.

6. Edge, M.; Allen N.S.; Williams, D.A.R.; Thompson, F. Polymer Degradation and Stability **1992,** *35,* 147-155.

7. Jacobsen, M. In *Polymers in Conservation*; Allen, N.S.; Edge, M.; Horie, C.V., Eds.; The Royal Society of Chemistry: Cambridge, England, 1992; pp. 151-158.

8. Shinagawa, Y.; Murayama, M.; Sakaino, Y. In *Polymers in Conservation*; Allen, N.S.; Edge, M.; Horie, C.V., Eds.; The Royal Society of Chemistry: Cambridge, England, 1992; pp. 138-150.

9. Byrk, M.T. In *Degradation of Filled Polymers: High Temperature and Thermal-Oxidative Processes;* Kemp, T.J.; Kennedy, J.F., Eds.; Ellis Horwood Ltd: Chichester, England, 1991.

10. Edwards, H.G.M.; Johnson, A.F.; Lewis, I.R.; Turner, P. Polymer Degradation and Stability **1993,** *41,* 257-264.

11. Brown, T.; Dronsfield, A.; Cheetman, C.; Cope, B.; Matthews, A.; Maddock, D., Internal Paper, University of Derby, Derby, England, 1998.

12. Cardamone, J.M.; Keister, K.; Osarch, A.H. In *Polymers in Conservation*; Allen, N.S.; Edge, M.; Horie, C.V., Eds.; The Royal Society of Chemistry: Cambridge, England, 1992; pp. 108-124.

13. Pullen, D.; Heuman, J. *Preprints of Contributions to the Modern Organic Materials Meeting;* University of Edinburgh, Edinburgh, Scotland, 1988; pp. 55-66.

14. DeCroes, G.C.; Tamblyn, J.W.; *Modern Plastics, 1952,* pp. 29-127.

15. Stannett, V. *Cellulose Acetate Plastics;* Temple Press Ltd: London, 1950; pp. 43-75.

Chapter 13

Spectroscopic Investigation of the Degradation of Vulcanized Natural Rubber Museum Artifacts

Sandra A. Connors[1,4], Alison Murray[1], Ralph M. Paroli[2], Ana H. Delgado[2], and Jayne D. Irwin[2]

[1]Art Conservation Program, Department of Art, Queen's University, Kingston, Ontario K7L 3N6, Canada
[2]National Research Council of Canada, Institute for Research in Construction, Building Envelope and Structure, Ottawa, Ontario K1A 0R6, Canada

Vulcanized natural rubber artifacts degrade severely over time, making the evaluation of these artifacts very difficult. This study, therefore, chemically characterized artificially and naturally aged vulcanized natural rubber samples using three Fourier-transform infrared spectroscopy techniques in order to determine the usefulness of each technique for the evaluation of this material. Physical characterization of vulcanized natural rubber was done using scanning electron microscopy as well as mechanical testing.

Chemical analysis indicated that oxidative degradation of the material had occurred due to accelerated aging of the samples. The physical changes occurring in the artificially aged samples closely resembled the changes found in many museum artifacts. These physical changes included: cracking, hardening, and embrittlement. There was, however, no specific structural change occurring in the material that was found to be linked to a specific physical sign of degradation.

---

[4] Current address: Research Center on the Materials of the Artist and Conservator, Carnegie Mellon Research Institute, 700 Technology Drive, Suite 3210, Pittsburgh, PA 15230.

Introduction

Artifacts made of polymeric materials can be found in almost every museum throughout the world. Machinery, household goods, medical equipment, as well as fine art objects, have all been made, at least in part, from polymers.[1,2,4] These collections are the concern of conservators and conservation scientists who have to make decisions regarding their exhibition, storage, and care. Most artifacts made of polymeric materials are difficult to conserve and characterize, but vulcanized natural rubber artifacts pose a particular problem because of the severity of degradation frequently found in these artifacts. This degradation can manifest itself as bloom, random or linear cracking patterns, hardening and brittleness or softening and tackiness.[4] These physical changes are visible to the human eye. It is essential, however, to have a thorough understanding of the chemical and physical changes that have occurred before attempting conservation, as treatments done to vulcanized natural rubber artifacts may cause damage and accelerate the rate of deterioration. Previous research in this area has provided possible mechanisms for the degradation of vulcanized natural rubber.[3-6] This research, therefore, focuses on studying the chemical and physical changes occurring in vulcanized natural rubber as a result of degradation and evaluating specific instruments for their usefulness in the characterization of vulcanized natural rubber.

Fourier-transform infrared (FTIR) spectroscopy techniques have frequently been used to evaluate elastomeric materials.[7-9] Three FTIR techniques were evaluated for their usefulness in characterization of vulcanized natural rubber, which had been either naturally or artificially aged. These techniques include: attenuated total reflectance-microscopy-Fourier-transform infrared spectroscopy (ATR-microscopy-FTIR), attenuated total reflectance-Fourier-transform infrared spectroscopy (ATR-FTIR), and photoacoustic-Fourier-transform infrared spectroscopy (PAS-FTIR). Scanning electron microscopy (SEM) and mechanical testing were used to evaluate changes in the morphology and tensile properties of the artificially aged samples.

Experimental

Samples

Vulcanized natural rubber surrogate samples were prepared by the Akron Rubber Development Laboratory, Inc. in Akron, Ohio, using ASTM standard D 3184 formula 2A as a guideline. The mixing and vulcanization procedures conform to ASTM standard D 3182 with a cure temperature of 284°F and a curing time of 40 minutes.

Naturally aged vulcanized natural rubber samples were obtained from artifacts in the collection at the Henry Ford Museum & Greenfield Village, Dearborn, Michigan. When possible, pieces of rubber which had fallen off the object were used. If no detached pieces were available, a sample was cut, using a scalpel, from an already damaged area or a portion which could not be seen when the artifact was exhibited. Samples were obtained from artifacts with varying degrees of visible degradation. Samples taken from artifacts are listed in Table I.

Table I. Sampled Henry Ford Museum & Greenfield Village Artifacts.

Henry Ford Museum & Greenfield Village Artifacts	Visible Signs Of Degradation
c. 1880 Drum Cylinder Press	• severe cracking in rubber • softness and tackiness
1907 Harley Davidson tire	• severe cracking in rubber • hardness and brittleness • pieces deadhering from the substrate of the tire
c. 1910 washing machine	• fine cracking in rubber • some flexibility, though some hardening may have taken place

Accelerated Aging

The vulcanized natural rubber surrogate samples were artificially aged using a Q-panel QUV. Forty-eight specimens of vulcanized natural rubber with dimensions 7 cm x 14 cm were cut out of the original 144 cm x 144 cm panels received from Akron Rubber Development Laboratory. Six of the surrogate samples were retained as unexposed, control samples. The remaining samples were put into the QUV. UV-A fluorescent bulbs (40W) were used that had a peak emission at 340 nm. These bulbs were chosen over UV-A bulbs with a peak emission at 351 nm because of their similarity to outdoor weathering conditions. Many historic rubber artifacts have had a history of outdoor use prior to exhibition in a museum (i.e. automobiles and agricultural equipment). The bulbs and samples were rotated, according to ASTM standard G 53, every 400 and 168 hours, respectively. Six samples were removed after one week, one month, two months, three months and four months of QUV exposure. At each time interval, chemical and physical analyses were performed on the samples.

Attenuated Total Reflectance-Microscopy-Fourier-Transform Infrared Spectroscopy (ATR-Microscopy-FTIR)

Five specimens from each sample were analyzed using a Nicolet Nic-Plan microscope equipped with an MCT-A detector. No preparation was needed prior to analysis of the specimens. Spectra were collected using an attenuated total reflectance (ATR) attachment with a 45° germanium (Ge) crystal. Each specimen was placed on the microscope slide attached to a contact alarm. A 10X-glass objective was used to select the area to be sampled and to focus the microscope. Once in focus, contact was made between the ATR objective and the specimen. The collection parameters were: 1.5 mm aperture, 500 scans, mirror velocity set to 55 (3.16 cm/sec), 8 cm^{-1} resolution and Happ-Genzel apodization. The contact alarm was kept at 2-4 mA when the specimens were being scanned to ensure good contact between the specimen and the germanium crystal. The ATR and baseline correction routines from the Omnic software (Nicolet Instruments) were applied to all spectra before plotting.

Avatar System 360 with OMNI-Sampler Attachment (OMNI-Sampler)

Five specimens (one from each sample after zero, one month, two months, three months, and four months of QUV exposure) along with one naturally aged museum sample were analyzed using an Avatar System 360 with Omni-Sampler attachment. The Omni-Sampler was equipped with a Ge crystal for ATR. After each specimen was placed on top of the germanium crystal, pressure was applied by tightening the pressure tower, to create good contact between the specimen and the crystal. The collection parameters were: 50 scans, 0.6329 cm/sec mirror velocity, 4 cm^{-1} resolution, and Happ-Genzel apodization. ATR and baseline correction routines from the Omnic software (Nicolet Instruments) were applied to all spectra before plotting.

Photoacoustic-Fourier-Transform Infrared Spectroscopy (PAS-FTIR)

Five trial specimens from each sample were cut using a hole punch. Each specimen was approximately 6 mm in diameter and approximately 2 mm in thickness. It was necessary to cut each specimen in half, widthwise, using a utility knife to reduce the thickness to approximately 1 mm. Analysis of the specimens was performed using a Nicolet 800 infrared spectrometer equipped with a photoacoustic cell and detector (MTEC Photoacoustics model 200). Each specimen was placed in the sample pan with the newly exposed side of the sample facing upward. The pan was introduced into the cell and purged with helium gas (5 ml/sec) for 1-2 minutes. The specimens were scanned using the following experimental parameters: 8 cm^{-1} resolution, 300 scans, and mirror velocity set to 20 (0.32 cm/sec). The sample spectra were ratioed against a carbon black standard reference membrane background collected the same day. Omnic software from Nicolet Instruments was used to correct for the non-linearity of the PAS detector.

Scanning Electron Microscopy (SEM)

A JEOL JSM-T300 scanning electron microscope was used to obtain morphological information about the samples. Specimens approximately 1 cm square were cut from unexposed and exposed samples. Each specimen was attached to a mount using double-sided tape and a Hummer VI-A Sputtering System was used to coat the specimens with gold. The fracture point of the tensile strength specimens was also examined using the SEM. These specimens were mounted using hot glue and coated with gold using the same sputtering apparatus. Micrographs were obtained for each specimen.

Tensile Data

Ten dog-bone specimens were cut for each sample with die C, as specified in ASTM D 412, using a Neaf hydraulic punch press. The thickness of each specimen was precisely measured, taking an average of five points measured along the length of the specimen using a Mitutoyo digital micrometer.

The samples were tested using an Instron 4502 Automated Materials Testing System with Series IX software. The room was kept at a constant temperature of 23 ± 2°C and a relative humidity of 50 ± 5%. The specimens were tested at a speed of 500 mm/min using a 1 kN capacity load cell. Pneumatic grips were used to hold the specimens. A sampling rate of 10 pts/sec and a gauge length of 25 mm were used. An Instron XL Extensiometer was used to measure displacement.

Results and Discussion

Figures 1-3 display spectra obtained using ATR-microscopy-FTIR. Figure 1 shows the spectrum of the unexposed vulcanized natural rubber sample. Figures 2 and 3 display spectra of the artificially aged and museum sample respectively. The spectra in Figures 1 and 2 contain peaks which are typical of cis-1,4-polyisoprene, the base polymer of natural rubber. These include strong peaks in the region from 3000-2800 cm^{-1} (aliphatic CH_2 and CH_3 stretch), a weak, broad peak between 1700-1500 cm^{-1} (C=C stretch), two strong peaks between 1450 cm^{-1} (CH and CH_2 deformation) and 1370 cm^{-1} (CH_3 deformation), and a strong peak around 830 cm^{-1} (R_2C=CHR).[10-11] The peak at 1539 cm^{-1}, present in Figure 2, was also present in spectra of some of the unexposed vulcanized natural rubber samples. The peak, however, did not consistently appear. This suggests that it may be the result of a processing agent or additive which is not mixed uniformly through the polymer system.

ATR-microscopy-FTIR has only shown limited usefulness for the evaluation of the surrogate and naturally aged samples as two problems arose. The first is the high level of carbon black in the surrogate samples. Carbon black is used in vulcanized natural rubber formulations as filler and to provide some protection against photo-degradation. Because of its strong absorbing power in the infrared region, carbon black masks functional group information about the polymer system. Due to this interference, the spectra of material containing carbon black tend to be noisy and have peaks of low intensity. As material containing carbon black ages, the carbon black migrates to the surface. This compounds the problem of spectral interference by masking changes which might indicate the presence of degradation products. The

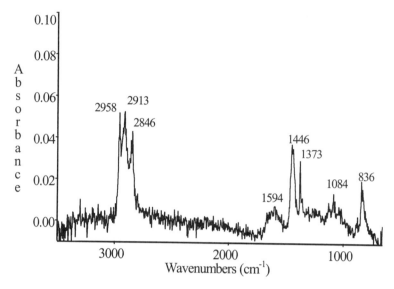

Figure 1. ATR-Microscopy-FTIR Spectrum of an Unexposed Vulcanized Natural Rubber Surrogate Sample.

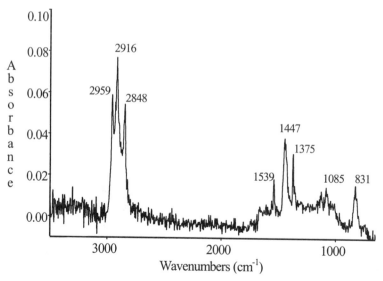

Figure 2. ATR-Microscopy-FTIR Spectrum of a Vulcanized Natural Rubber Surrogate Sample after One Week of QUV Exposure.

second problem with using the ATR-microscopy-FTIR arises from the fact that many samples become hardened and embrittled as a result of aging. When the IR-ATR objective is placed in contact with the sample, the sample material must be soft enough to allow good contact with only a small amount of pressure. The spectrum in Figure 3 is from a naturally aged museum sample (c. 1910 washing machine) and it is typical of the spectra obtained from museum samples using this technique. The only peaks present (2362 cm^{-1} and 2342 cm^{-1}) are due to carbon dioxide in the surrounding atmosphere. This indicates that good contact was not made between the crystal and sample.

Figures 4-7 display spectra obtained using the Avatar System 360 with OMNI-Sampler attachment. All spectra show peaks which are typical of cis-1,4-polyisoprene. The peak near 1540 cm^{-1}, present in the ATR-microscopy-FTIR spectrum, is also present here. In comparison to the ATR-microscopy-FTIR spectra of exposed samples, however, the OMNI-Sampler spectra give a much better indication of degradation of the vulcanized natural rubber. After two months of exposure (Figure 5), there is already a noticeable decrease in the intensity of peaks between 3000 cm^{-1} and 2800 cm^{-1} (CH$_2$ and CH$_3$ stretch) as well as a decrease in peaks between 1450 cm^{-1} (CH and CH$_2$ deformation) and 1370 cm^{-1} (CH$_3$ deformation). After four months of exposure (Figure 6), the peaks around 2900 cm^{-1} (CH$_2$ and CH$_3$ stretch) are no longer noticeable. In both Figures 5 and 6, there appears to be a broadening of the peak at 1638 cm^{-1} to include a greater intensity at 1707 cm^{-1}. This may indicate the presence of oxidative degradation occurring in the material. Figure 7 shows the spectrum of a c. 1880 drum cylinder press sample. All the peaks typical of cis-1,4-polyisoprene are present in the spectrum. The peaks near 3400 cm^{-1} and 1650 cm^{-1} may be due to absorbed water in the sample.

The two problems that arose when evaluating the naturally aged and artificially aged samples using ATR-microscopy-FTIR were not major factors in this second technique. First the area sampled with the OMNI-Sampler attachment (2 mm) is larger than the area sampled by the ATR-microscopy-FTIR instrumentation. The larger area being sampled minimizes the interference caused by the carbon black in the sample. It does, however, run the risk of having the sample not be representative of degradation occurring throughout the object. The second problem in using the ATR-microscopy-FTIR instrument was the small amount of pressure which could be applied to the germanium crystal. When the sample is placed on the crystal, a pressure tower is used to create uniform pressure on the sample by pressing it against the crystal. A slip-clutch mechanism on the pressure tower allows the maximum amount of pressure the crystal can withstand to be achieved each time the attachment is used. It is possible, therefore, to obtain better contact between the crystal and sample even when the sample has become somewhat cracked and brittle. This technique, however, does require that the sample be able to yield slightly under pressure. For significantly hardened samples, the OMNI-Sampler would not be able to obtain good contact between the sample and Ge crystal.

The results obtained using PAS-FTIR show varying degrees of success. The spectra of surrogate samples initially showed great promise, as all functional groups typical of cis-1,4-polyisoprene were present in the spectrum of the unexposed surrogate sample (Figure 8). The peaks at 1312 cm^{-1}, 1217 cm^{-1}, and 987 cm^{-1} may be

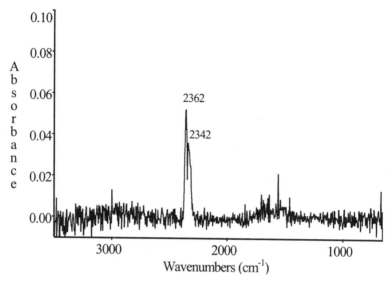

Figure 3. Attempt of an ATR-Microscopy-FTIR Spectrum of a c.1910 Washing Machine Sample.

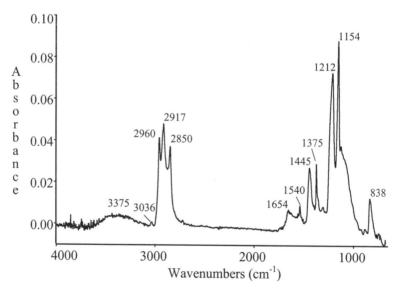

Figure 4. OMNI-Sampler Spectrum of Unexposed Vulcanized Natural Rubber.

174

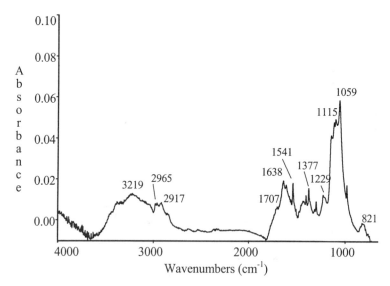

Figure 5. OMNI-Sampler Spectrum of Vulcanized Natural Rubber after Two Months of QUV Exposure.

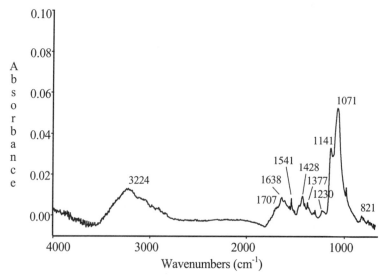

Figure 6. OMNI-Sampler Spectrum of Vulcanized Natural Rubber after Four Months of QUV Exposure.

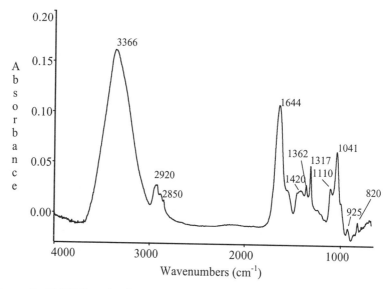

Figure 7. OMNI-Sampler Spectrum of a c. 1880 Drum Cylinder Press Sample.

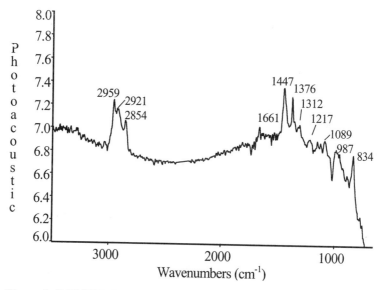

Figure 8. PAS-FTIR Spectrum of Unexposed Vulcanized Natural Rubber.

peaks, although it is difficult to obtain a conclusive identification with such a high baseline in the spectrum. The high baseline in this and other PAS-FTIR spectra of surrogate samples exposed for two and four months (Figures 9 and 10, respectively) results from the high level of carbon black in the polymer system. As the carbon black migrates to the surface, the amount of IR signal absorbed by the carbon black increases. The baseline, therefore, will increase as the material ages until the peaks (due to cis-1,4-polyisoprene) are masked and eventually obscured by the carbon black. This problem is intensified by the fact that the background is scanned against a carbon black reference membrane, which adds to the high baseline of the spectra.

There appears to be some change in the functional groups of the surrogate sample spectrum, after only two months of QUV exposure (Figure 9). The peak at 1661 cm^{-1} is most likely due to the stretching motion of the carbon, carbon double bond. This peak, however, seems to have broadened as a result of QUV exposure. The peak at 1661 cm^{-1} diminishes and a peak around 1710 cm^{-1} appears, compared to the PAS-FTIR spectrum of the unexposed sample (Figure 8). This increase in intensity near 1710 cm^{-1} may be the result of oxidative degradation of the vulcanized natural rubber. An increase in this area would indicate an increase in the amount of carbonyl-containing groups which would be consistent with the oxidative degradation of vulcanized natural rubber described elsewhere.[4-6,12,13] After four months of exposure (Figure 10), however, almost all functional group information is obscured. The peaks around 2959 cm^{-1} (CH$_2$ and CH$_3$ stretch) are no longer visible. The baseline of the spectrum has increased significantly, making it very difficult to distinguish any of the peaks from the background with absolute certainty.

Naturally aged museum samples were also evaluated using PAS-FTIR. Most of the spectra of these samples have well defined peaks and a low signal-to-noise ratio. Figure 11 shows the PAS-FTIR spectrum of a c. 1910 washing machine sample. The baseline in this spectrum is higher than would be expected because of the carbon black in the polymer system; however, functional group information is present. The lower baseline in Figure 11 may be due to a lower level of carbon black in the museum sample formulations compared to that of the surrogate samples. All peaks typical of cis-1,4-polyisoprene are present along with some peaks which indicate the presence of degradation products. The spectrum of the c. 1910 washing machine sample exhibits two peaks (3388 cm^{-1} and 1668 cm^{-1}), which may be due to absorbed water in the sample. There is also a peak at 1712 cm^{-1}, which is most likely due to degradation products that contain carbonyl groups. It is also possible that the peaks at 3388 cm^{-1}, 1668 cm^{-1} and 1712 cm^{-1} may be due to other formulation components not present in the surrogate sample formulation.

Figures 12-14 show the changes in appearance that occurred in the vulcanized natural rubber material as a result of accelerated aging. The most prominent change is the random cracking pattern on the surface of the material. The main cracks in this pattern meet one another to form polygon shapes which are relatively uniform in size; smaller cracks do branch off the main cracks which results in a similar cracking pattern within each polygon shape. These cracks propagate down into the center of the rubber material rather than along the surface. As a result, the tops of the cracks (at the surface of the material) widen, creating a V-shape, whose base is toward the center of the material. The edges of the cracks appear very jagged (see Figure 14).

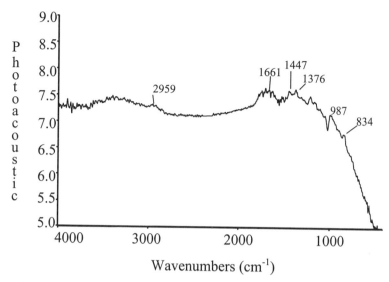

Figure 9. PAS-FTIR Spectrum of Vulcanized Natural Rubber after Two Months of QUV Exposure.

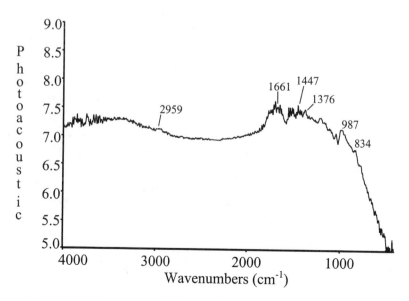

Figure 10. PAS-FTIR Spectrum of Vulcanized Natural Rubber after Four Months of QUV Exposure.

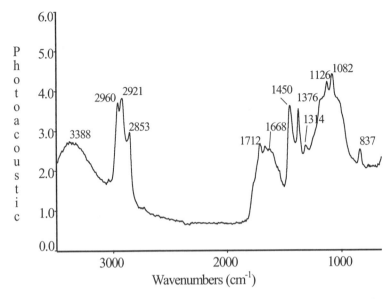

Figure 11. PAS-FTIR Spectrum of a c. 1910 Washing Machine Sample.

Figure 12. SEM Image of Unexposed Vulcanized Natural Rubber. (200X Magnification).

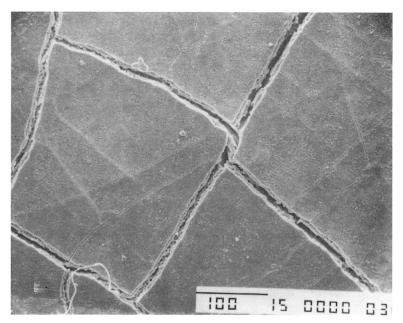

Figure 13. SEM Image of Vulcanized Natural Rubber after Two Months of QUV Exposure. (200X Magnification).

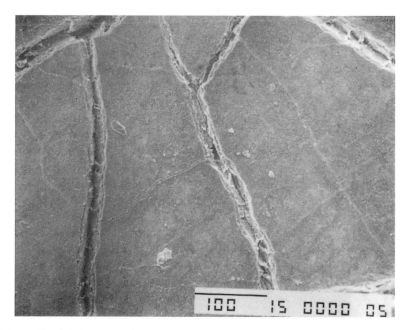

Figure 14. SEM Image of Vulcanized Natural Rubber after Four Months of QUV Exposure. (200X Magnification).

The cracking found on the tires of a 1907 Harley Davidson motorcycle (Figure 15) is similar to that found on the exposed surrogate samples. The cracks on the Harley Davidson tire also show the formation of polygon shapes in the rubber.

The rubber in the QUV-exposed samples also exhibits an increase in brittleness. This is indicated by the changes occurring in the appearance of fracture points as a result of aging (see Figures 16-18). In Figure 16, the fracture of the unexposed vulcanized natural rubber sample is torn and uneven, and begins at the upper corner of the sample (the upper left corner of the image). The mirror (semi-circular) shape, of the fracture, typical of a more brittle fracture, appears after two months of exposure and becomes more prominent with increased QUV exposure. The fracture lines formed become straighter and more defined. The most common fracture point for these specimens was a corner, as seen in Figure 16; however, on occasion the fracture point was in the center of the specimen as seen in Figure 18.

Evaluation of tensile properties of the surrogate samples has provided information about the changes in the physical properties of this material as a result of accelerated aging. Figure 19 shows the percent elongation of the vulcanized natural rubber as a result of QUV exposure. From this chart it is evident that the percent elongation decreases with increased exposure in the QUV. This indicates that the vulcanized natural rubber specimens become brittle as a result of QUV exposure.

Conclusions

This study has provided the authors with information regarding the physical and chemical changes in vulcanized natural rubber as a result of accelerated and natural aging. The diagnostic tools evaluated during this study have shown varied degrees of usefulness for evaluating the degradation of vulcanized natural rubber.

The Avatar System 360 with OMNI-Sampler (ATR) attachment appears to be a more reliable tool for evaluating the vulcanized natural rubber samples. The pressure tower on this attachment was able to consistently provide enough pressure on the sample to achieve good contact without damaging the crystal, even when the surface of the material was somewhat uneven and cracked. It was not, however, able to provide good contact when the sample was severely degraded (i.e. very hard and brittle).

PAS-FTIR was found to be very useful for evaluating the severely degraded museum samples, which had become hard and brittle with age. The museum sample spectra exhibited good resolution of the peaks and good signal-to-noise ratio, regardless of the hardness and unevenness of the sample surface. The quality of the PAS-FTIR spectra was, however, dependent on the carbon black content of the material. The naturally aged museum samples appear to have less carbon black, because of the significantly better quality of the spectra obtained, compared to the surrogate samples; however, this was not yet confirmed with chemical analysis to determine the carbon black loading in each sample.

ATR-microscopy-FTIR showed only a minor ability to evaluate the vulcanized natural rubber samples. The inability of the germanium ATR crystal to withstand pressure prevented the achievement of good contact between the crystal and the

Figure 15. Detail of Tire from a 1907 Harley Davidson Motorcycle.

Figure 16. SEM Image of Fracture Point of an Unexposed Vulcanized Natural Rubber Specimen. (50X Magnification).

Figure 17. SEM Image of Fracture Point of a Vulcanized Natural Rubber Specimen after Two Months of QUV Exposure. (50X Magnification).

Figure 18. SEM Image of Fracture Point of a Vulcanized Natural Rubber Specimen after Four Months of QUV Exposure. (50X Magnification).

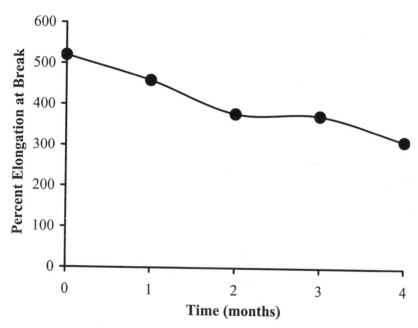

Figure 19. Changes in Percent Elongation of Vulcanized Natural Rubber Specimens as a Function of QUV Exposure.

sample and, consequently, prevented reproducible spectra from being obtained. The high carbon black content of the vulcanized natural rubber also compounded the problem by producing very noisy spectra with low peak intensity.

Accelerated aging caused significant physical changes in vulcanized natural rubber. Samples became harder and more brittle with increased QUV exposure, as was indicated by the decrease in percent elongation of the material as well as the changes in the appearance of the fracture point of vulcanized natural rubber specimens. The random cracking pattern that developed as a result of QUV exposure is typical of the cracking found on museum artifacts that have become hard and brittle over time.

Further research using the Avatar OMNI-Sampler attachment would be useful. This technique shows promise for the evaluation of minute samples and materials containing carbon black. Accelerated aging of vulcanized natural rubber samples should also be done using UV-A 351 bulbs in order to evaluate the degradation of rubber artifacts that were made solely for museum exhibition (i.e. fine art sculptures). Experimental protocols also need to be developed for quantifying the degradation of the vulcanized natural rubber material.

References

1. Allington, C. *Preprints of Contributions to the Modern Organic Materials Meeting Held at the University of Edinburgh, 14 & 15 April 1988*; SSCR Publications: Edinburgh, UK, 1988; pp 123-132.
2. Katz, S. *Early Plastics*; Shire Publications, Ltd.: Aylesbury, UK, 1986; pp 14-17.
3. Yano, S. *Rubber Reviews*. 1966, vol 39, 537-552.
4. Loadman, M. J. R. *Symposium '91-Saving the 20th Century: The Conservation of Modern Materials, 15 to 20 September 1991*; Canadian Conservation Institute: Ottawa, ON, 1993; pp 59-80.
5. McCord, M.; Daniels, V. *Preprints of Contributions to the Modern Organic Materials Meeting Held at the University of Edinburgh, 14 & 15 April 1988*; SSCR Publications: Edinburgh, UK, 1988; pp. 133-141.
6. Grattan, D. *J.IIC-CG.* 1978, vol 4, no 1, 17-26.
7. Chen, X.; Zhang, S.; Wang, X.; Yao, X.; Chen, J.; Zhou, C. *J. Appl. Polym. Sci.* 1995, vol 58, no 8, 1401-1405.
8. Thames, S.F.; Gupta, S. *J. Appl. Polym. Sci.* 1997, vol 63, no 8, 1077-1089.
9. Gaboury, S. R.; Urban, M. W. *Langmuir* 1994, vol 10, no 7, 2289-2293.
10. Colthup, N. B.; Daly, L. H.; Wiberley, S. E. *Introduction to Infrared and Raman Spectroscopy, 3rd ed.*; Academic Press: Boston, MA, 1990; pp 215-233, 247-259, 289-325.
11. Silverstein, R. M.; Bassler, G. C.; Morrill, T.C. *Spectroscopic Identification of Organic Compounds, 5th ed.*; John Wiley & Sons, Inc.: New York, NY, 1991; pp 91-102.
12. Kauffman, G. B.; Seymour, R. B. *J. Chem. Ed.* 1990, vol 67, no 5, 422-425.
13. Schidrowitz, P.; Dawson, T. R. *History of the Rubber Industry*; W. Heffer and Sons, Ltd.: Cambridge, UK, 1952; pp 1-63.

Chapter 14

Pyroxyline Paintings by Siqueiros: Visual and Analytical Examination of His Painting Techniques

Celina Contreras de Berenfeld[1], Alison Murray[2], Kate Helwig[3], and Barbara Keyser[2]

[1]Art History Program and [2]Art Conservation Program, Department of Art, Queen's University, Kingston, Ontario K7L 3N6, Canada [3]Canadian Conservation Institute, 1030 Innes Road, Ottawa, Ontario K1A 0M5, Canada

The role of David Alfaro Siqueiros (1896-1974) and his use of pyroxyline-based paints are crucial in the transition from oil to acrylic paints. During the early 20th century, his experimentation with this innovative paint medium had a great impact on North American traditional painting techniques (1). Although Siqueiros's paintings labeled as 'pyroxyline-based paintings' are highly prone to decay, little scientific research has been actually done to verify the presence of a pyroxyline-based medium (2). Identification of this type of paint film is usually based on visual examination. Consequently, pyroxyline paintings have often been confused with other painting media. In order to obtain data on Siqueiros's pyroxyline-based paintings, five of his paintings that records cite were made with pyroxyline-based paints were analysed using Fourier transform infrared spectroscopy and light microscopy. Results indicate that visual characteristics assumed for pyroxylines are a misleading guide in the labeling of paintings. Only two of the five paint films analysed were pyroxyline-based paints, two were painted in oil and one was an acrylic. The poor conditions of paint films are more probably caused by Siqueiros's unorthodox painting technique rather that his use of pyroxyline-based paint.

Introduction

David Alfaro Siqueiros (1896-1974) was a Mexican artist whose easel paintings and murals showed his sympathy with oppressed minorities around the world. His innovative imagery and his unconventional materials conveyed his rejection of what he believed were the tools that maintained class differences and injustices. For Siqueiros, it was the artist's duty to participate in the battle for democracy by experimenting with new techniques and new materials. Classical techniques should be rejected because they had emerged from despotic forms of society (3).

Siqueiros and his Use of Pyroxyline Paint Medium

Siqueiros spread his political and artistic message in extensive travels throughout Latin-America, Europe, and the United States. In 1936, under the auspices of the Communist party, he organized the Artists' Experimental Workshop Laboratory of Modern Techniques in Art in New York City (4). At the workshop, Siqueiros encouraged participants, who included Jackson Pollock and José Gutiérrez, to experiment with modern materials and new methods of application (5). Although the workshop was disbanded soon after he joined the Republican Army in Spain, the unorthodox methods used in the workshop had a profound impact on its participants. Siqueiros's interaction with Pollock is well documented and may have influenced Pollock's technical procedures (6). The relationship that developed between Siqueiros and Gutiérrez probably led to the establishment in 1945 of the Instituto de Ensayo de Pintura Materiales Plásticos at the National Polytechnic Institute in Mexico City. Gutiérrez advocated the use of vinylic and acrylic paints for artistic purposes. In a series of lectures, he encouraged artists throughout North America to use pyroxyline-based paints and other modern materials (7).

Siqueiros began to use pyroxyline-based paints in 1933, that been available in a range of consistencies since 1917 (8-11). Although Siqueiros has been credited as the first to use this material, Lodge has suggested that Max Ernst may have used it eight years earlier in his painting entitled *The Forest* (12).

Pyroxyline-based paints available in the 1930s were plasticed with camphor, aldol or castor oil (13). They were commonly marketed as varnishes, automotive paint, house paint and metal coatings (14,15). Although the instability of pyroxyline-based paints is now recognized, opinion on the lasting effects of pyroxylines was still divided in the 1940s and 1950s. Some chemists believed that pyroxyline-based paints readily discoloured, while others considered pyroxylines to be highly resistant to degradation (16-18). Siqueiros himself was concerned with the preservation of his work and was interested in making his own paints to improve the color and durability of the paints. He consulted managers at the Nitro Valspar Valentine Company and then the Dupont de Nemours Power Company on how to produce a matte finish on dry pyroxyline film (19). Siqueiros's attempts to obtain the formula for pyroxyline-based paints were unsuccessful. He instead experimented by adding different plasticizers to his pyroxyline-based paints and then modified them further with different solvents (20). It is not yet known if Siqueiros's experimentation with the paint media had a long-term effect of the degradation of his artwork.

Degradation of Pyroxyline

Pyroxyline is a polynitrate ester of cellulose with an average of 2.3 nitrate sidegroups (21). Although pyroxyline is moderately stable at temperatures above 100°C, if the material is exposed to ultraviolet radiation, high relative humidity, and specific air pollutants, the chemical structure will be prone to suffer acid-catalysed ester cleavage, scission of nitrogen-oxygen and ring disintegration. Metals such as calcium, silver, tin, iron, copper and zinc that frequently occur in pigments will react with pyroxyline (22). The type of degradation the material undergoes also depends on the plasticiser used. If it is volatile, for example camphor, the material can undergo stresses when the fumes are released, causing the formation of cracks. If the plasticiser is not volatile, such as castor oil, the plasticiser can migrate to the surface forming a tacky film.

Detrimental effects due to the exposure of ultraviolet radiation cause chain disintegration and the breaking of nitrogen-oxygen bonds. When the degradation process takes place, there is a loss of nitrogen and the backbone structure of the chain is broken. When this occurs, the material turns yellow to dark brown and becomes very brittle. Cracks may also form in this process as a way to release stresses. These reactions darken the paint film and create a brittle material (23). Investigations done by Stewart et al. and Derrick et al. give further insight into deterioration mechanisms of cellulose nitrate and the role that additives play in the degradation processes (24,25).

Siqueiros's use of pyroxyline is frequently cited; however, little analysis has been done to determine which paintings are actually executed in this medium. Although Fourier transform infrared spectroscopy has proved to be useful in the identification of modern media (26,27), sophisticated instrumental analysis is not always easily available in smaller museums. For this reason, identification of the media has frequently been based on visual characteristics, expected degradation effects, historic bibliographic sources and testimony of people who were close to Siqueiros while he was alive. Consequently, inconsistencies are found in publications about Siqueiros and on labels of his paintings shown in exhibitions. Same paintings have variously been described as oil, acrylic or pyroxyline (28-30). Observations made about the degradation of pyroxyline-based paintings also seem contradictory. Mateos noticed an equal degradation in the pyroxyline palette while Cuevas remarked that light colours, particularly white, are more prone to decay (31,32). To avoid jeopardizing the physical integrity of Siqueiros's paintings, it is imperative to identify their materials correctly.

Experimental

In order to determine more accurately Siqueiros's active use of pyroxyline-based paints, two samples were taken from five paintings by Siqueiros: *Mine Drillers* (c.1926), *Portrait of a Dead Child* (c.1931), *Peon* (c.1935), *Desfallecimiento* (c.1937-39) and *The First Victim's Father at The Cananea Strike* (1961). Each of these paintings has been labeled as being executed in pyroxyline-based paints, probably because their visual characteristics fit into the description of the appearance of pyroxyline technique as well as, in some cases, the expected degradation of pyroxyline. The dates of the paintings do not always conform with 1933 when Siqueiros began to use pyroxylines; however, curators believe the inscribed dates should not be considered a reliable source of information as Siqueiros frequently re-dated some of his paintings.

The paint samples were analysed with Fourier transform infrared spectroscopy using a Bomem MB-100 spectrometer or a Spectra-Tech IR-Plan Research microscope interfaced to a Bomem MB-120 spectrometer. The samples were mounted on a low-pressure diamond anvil microsample cell for analysis. In some cases, solvent extractions were performed in order to separate components of the paint medium from pigment and fillers. Compounds were identified by comparing the infrared spectrum to published reference material. The transmittance spectra shown in Figures 1-5 are the sum of 200 scans collected at a resolution $4cm^{-1}$. Bands due to atmospheric carbon dioxide in the $2320-2375cm^{-1}$ region have been removed and in some cases the spectra have been baseline corrected. Unmounted paint samples and paint cross-sections of *Mine Drillers* and *Portrait of a Dead Child* were analysed with a Nikon S-Kt light microscope.

Results and Discussion

Fourier Transform Infrared Spectroscopy

Analysis by FTIR spectroscopy revealed that two of the five paintings contained a cellulose nitrate medium, two others contained drying oil, and one contained an acrylic medium; however, these results should be considered preliminary since only one or two samples from each painting were analysed. It is possible that different media were used in different areas of the paintings. Drying oil was identified as the medium of the brown paint from both *Mine Drillers* and *Portrait of a Dead Child.* Infrared spectrum of the brown paint from *Portrait of a Dead Child* is shown in Figure 1. The lower spectrum shows the brown paint with no pre-treatment. The broad carbonyl band, centred at 1720 cm^{-1}, and the C-H stretches at 2926 cm^{-1} and 2854 cm^{-1} are due to the oil medium. The other bands in the spectrum are due to inorganic components in the paint (gypsum, quartz and calcium carbonate). The upper spectrum in Figure 1 shows the brown paint after removal of the carbonate with dilute hydrochloric acid. In this spectrum, additional characteristic bands for oil at 1463, 1412 and 1380 cm^{-1} are visible. The medium of the white ground layers in both *Mine Drillers* and *Portrait of a Dead Child* was identified as a collagen-type protein such as animal glue. Figure 2 shows the infrared spectrum of the ground layer from *Mine Drillers* with characteristic amide bands at approximately 1652, 1541 and 1457 cm^{-1}.

The lower trace in Figure 3 shows the spectrum of a sample of red paint from *Peon*, containing characteristic band for cellulose nitrate at approximately 1655, 1280 and 840 cm^{-1}. The carbonyl band at 1724 cm^{-1} is due to the plasticizer. The upper trace shows the spectrum of the plasticizer extracted from the paint with chloroform. The infrared spectrum of the plasticizer shows strong absorptions in the C-H and carbonyl regions, combined with bands at 1600, 1580, 1280, 1123, 1073, 1041, and 743 cm^{-1}. This spectrum is indicative of an alkyl phthalate.

Figure 4 shows the infrared spectrum of an orange paint sample from *Desfallecimiento*. The paint was found to contain both cellulose nitrate and an ester-containing compound. Solvent extractions were unsuccessful in isolating the ester-containing component for identification. The compound could be a plasticizer for the cellulose nitrate, or it could be due to an oil mixed with the cellulose nitrate medium. Further analysis using a separation technique such as gas chromatography/mass chromatography could determine if a mixed medium was used in this painting.

The medium of the red paint in *The First Victim's Father at the Cananea Strike* was identified as an acrylic, probably a copolymer containing poly (ethyl methacrylate). The spectrum of the red paint is shown in Figure 5. The C-H stretching bands at 2984 and 2952 cm^{-1}, the carbonyl band at 1731 cm^{-1}, and the group of bands between 1480-1360 cm^{-1} are characteristic of an acrylic medium. The presence of bands at approximately 860 and 752 cm^{-1} suggests that poly(ethyl methacrylate) is present. Some of the bands of the acrylic medium in the 1200-900 cm^{-1} region are masked by absorptions of a kaolin-group clay that is also present in the paint.

Visual Examination and Light Microscopy

Mine Drillers (Figure 6) was executed on a canvas support. The paint film is dark and brittle. Light microscopy analysis of the cross-section (Figure 7a) indicates that in some areas the painting has a wax coating (a) that explains a semi-matte finish found in some sections of the painting. The cross-section also shows a dark brown paint layer (b) and a course, porous and water-sensitive ground layer (c). Another cross-section (Figure 7b) shows a thick ground layer (b) sandwiched between two grey paint films (a, c). The canvas support and the water-sensitive ground layer react to environmental changes especially where the ground layer is thick and porous. The ground layer has developed cracks and the surface is showing cleavage.

Portrait of a Dead Child (Figure 8) was painted with oil medium on canvas support. This explains the smooth surface and occasional *impasto*, characteristic of the traditional oil painting technique as well as the fact that the painting was lined with a

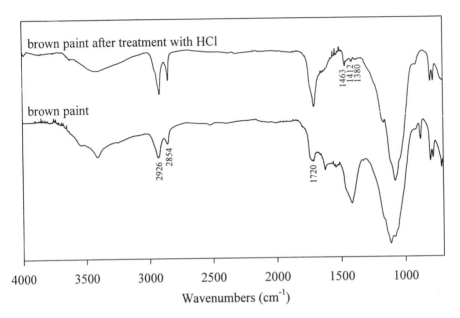

Figure 1. FTIR spectra of paint medium in "Portrait of a Dead Child".

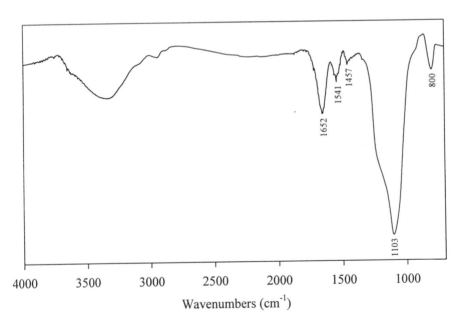

Figure 2. FTIR spectra of ground layer in "Mine Drillers".

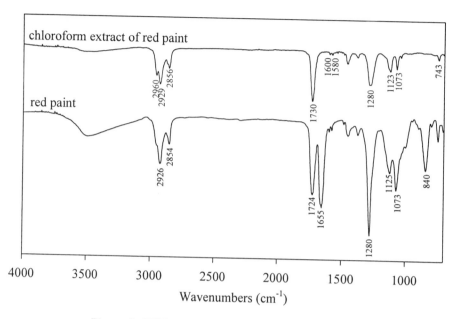

Figure 3. FTIR spectra of paint medium in "Peon".

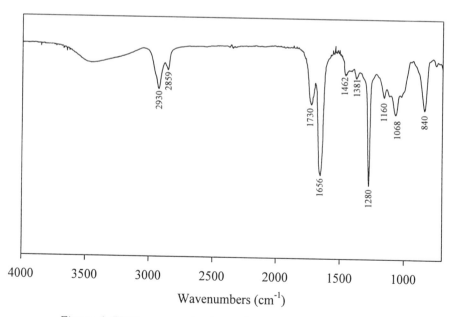

Figure 4. FTIR spectra of paint medium in "Desfallecimiento".

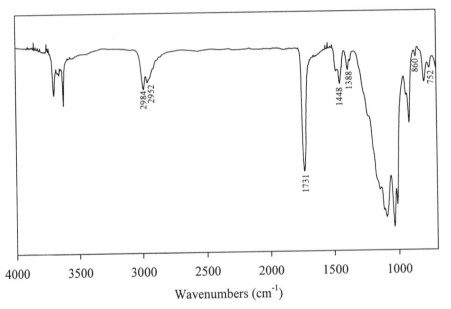

Figure 5. FTIR spectra of paint medium in "The First Victim's Father at Cananea Strike".

Figure 6. D.A Siqueiros. "Mine Drillers" c.1926. The Soumaya Museum, Mexico City. Photograph by Javier Hinojosa. (Reproduced with permission from the Soumaya Museum. Copyright 1996.)

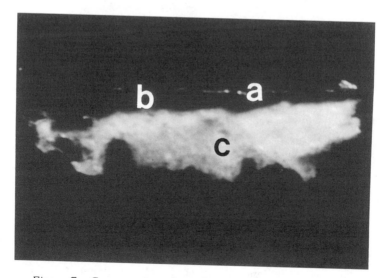

Figure 7a. Cross-section of sample no.1 from "Mine Drillers".

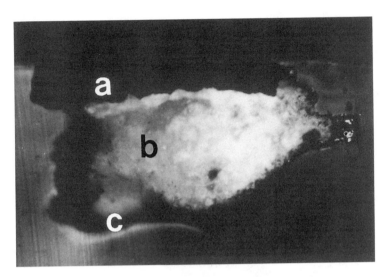

Figure 7b. Cross-section of sample no.2 from "Mine Drillers".

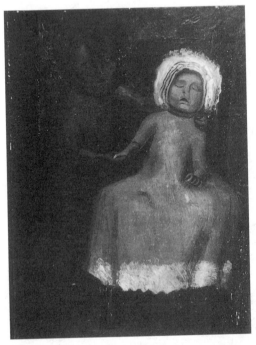

Figure 8. D.A Siqueiros. "Portrait of a Dead Child" c.1931. The Soumaya Museum, Mexico City. Photograph by Javier Hinojosa. The Soumaya Museum, Mexico City. Photograph by Javier Hinojosa. (Reproduced with permission from the Soumaya Museum. Copyright 1996.)

wax-resin adhesive, a treatment that requires heat and pressure. A comparison of the painting with a 1968 black and white photograph confirms that the oil film has darkened substantially over the last three decades, especially in the brown and orange areas (*33*). The pigments Siqueiros used are dark; however, the painting has gone through a wax relining treatment that has caused additional darkening of colours.

Results also indicate that the ground of *Portrait of a Dead Child* has similar characteristics to the ground of *Mine Drillers*. A cross-section (Figure 9a) shows that under the paint layer (a) there is an uneven ground layer (b). Another cross-section (Figure 9b) shows a thin paint layer at the top (a), resulting from the lining, and a very thick ground layer that encompasses a large dark particle (c) that looks similar to the painting material in the top layer.

The ground layers in both paintings are porous and water-soluble, except for areas impregnated with wax, consequently these are especially susceptible to mechanical stresses caused by environmental changes. It was an important discovery to find a ground layer sandwiched between two paint films (Figure 7a) and a black particle sandwiched between ground layers (Figure 9a). This indicates that Siqueiros's unorthodox technique of putting down multiple layers to create heavy textures, used in both paintings, developed mechanical crackle and cleavage in areas not embedded in wax. In both paintings, X-ray radiography confirmed that Siqueiros did not paint another image beneath the visible image.

Peon (Figure 10) was painted with a pyroxyline-based medium on a compressed-wood panel. The paint film has a waxy translucent appearance and it is also in good condition. It has a semi-textured surface and its appearance looks more similar to that of wax. On the other hand, the paint film of *Desfallecimiento* (Figure 11), also a cellulose nitrate paint media, showed brittleness and severe flaking in areas with lighter hues and in some areas near the edges of the cardboard support. The paint film is semi-glossy and translucent in appearance. Even though the paint layers have been thinly applied, isolated brushstrokes are clearly visible. The colours are intense and the image is well defined. It was painted on a white gesso ground, now brittle, that was applied to a cardboard support. The degradation of the pyroxyline, the brittle ground layer, or the flexible and acidic cardboard used as the support are causing the poor state of the painting. Both pyroxyline-based paintings share a translucent appearance.

The First Victim's Father at The Cananea Strike (Figure 12) was painted on a wood panel and the paint layer has an opaque semi-mat surface with a heavily textured surface. The brush strokes are noticeable and well defined which is characteristic of a fast drying medium. The painting has the aesthetic characteristics described for the pyroxyline technique; however, the painting medium has been identified as an acrylic. The painting is in a good state and the colours are still intense. The good state of preservation is probably related to the medium used and to its recent date of execution of 1961.

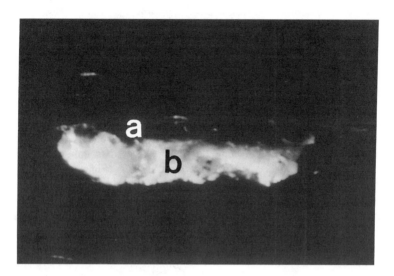

Figure 9a. Cross-section of sample no.1 from "Portrait of a Dead Child".

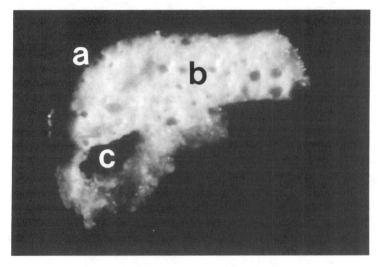

Figure 9b. Cross-sections of sample no.2 from "Portrait of a Dead Child".

Figure 10. D.A. Siqueiros. "Peon" c.1935. Photograph by Javier Hinojosa.

Figure 11 D.A. Siqueiros. "Desfallecimiento" 1937-39. The Soumaya Museum, Mexico City. Photograph by Javier Hinojosa. The Soumaya Museum, Mexico City. Photograph by Javier Hinojosa. (Reproduced with permission from the Soumaya Museum. Copyright 1996.)

Figure 12. D.A. Siqueiros. "The First Victim's Father at Cananea Strike" 1961. The Soumaya Museum, Mexico City. Photograph by Javier Hinojosa. The Soumaya Museum, Mexico City. Photograph by Javier Hinojosa. (Reproduced with permission from the Soumaya Museum. Copyright 1996.)

Conclusion

A medium cannot be determined based solely on visual appearances. The analysis of the five paintings indicates that the visual characteristics and degradation signs are linked to factors other than the presence of pyroxyline and its decay. Results from this research indicate that the dates inscribed have proven to be consistent with the type of media Siqueiros was using during certain periods. Although there is a general belief that he often mixed oil and pyroxyline-based paints, this research has yet to find evidence that he mixed different media in a single work. Further analysis would have to be done to confirm this.

Siqueiros's use of textured surfaces was a common practice before he started to use his pyroxyline technique. Texture in his early oil works was achieved by the use of thick ground layers, superimposing oil paint and ground layers, and also by adding fillers to the paint medium. The condition of paintings executed with pyroxylines varies greatly. Consequently, visual characteristics of pyroxylines cannot be generalised. In the two examples analysed, the paintings vary from having a thick, waxy surface, to having a thin, semi-matte surface. Although both pyroxyline paint films share a degree of translucency, this should not be considered standard in cellulose nitrate paintings. The degradation of Siqueros's painting executed in pyroxyline-base medium still need further research.

Acknowledgements

The authors would like to thank:

- The School of Graduate Studies at Queen's University, Kingston, Ontario and the Fondo Nacional para la Cultura y las Artes CNCA, Mexico City, for the financial support provided for this research.
- The Soumaya Museum, Carso A.C., Mexico City, for providing us with the samples of Siqueiros's paintings, for making their x-ray radiograph records available to us, and for allowing us to reproduce their photographic material conditionally.
- Historian Julieta García García from the Curatorial Department at the Viregal Art Museum of Puebla for her valuable advice on Siqueros's paintings.
- Dr. Cathleen Hoeniger and Dr. David McTavish from the Department of Art at Queen's University for their valuable advice on curatorial and institutional grounds.
- The Natural Sciences and Engineering Research Council of Canada for funding from an operating grant.

Literature Cited

1. Marontate, J. Ph.D. thesis, Université de Montréal, Montréal, QC, 1996; pp 34-69, 156-196.
2. Contreras M., C. M.A.C. thesis, Queen's University, Kingston, ON, 1998.
3. Contreras, C. In *Student Papers Twenty-Forth Annual AGPIC Student Conference*; Barbara Keyser, Ed.; Art Conservation Program, Queen's University: Kingston, ON, 1998, pp 75-86.
4. Hurlburt, L. In *The Mexican Muralists in the United States*; University of New Mexico Press: Albuquerque, NM, 1989; pp 238-245.
5. Herner, I. In *Otras rutas hacia Siqueiros*; Curare-Instituto Nacional de Bellas Artes: Mexico City, Mexico, 1996.
6. Medina, A. *Art Nexus* 1996, *vol 19*, pp 42-53.
7. National Gallery of Canada Archives, National Gallery of Canada fonds, 7.4G Outside Activities/Organizations, Gutiérrez, José (Lectures).
8. Herner, I. In *Otras rutas hacia Siqueiros*; Curare-Instituto Nacional de Bellas Artes: Mexico City, Mexico, 1996, p 182.
9. Carlisle, C. *Art News* 1996, *vol 95*, p 57.
10. Siqueiros, D. A. *Palabras de Siqueiros*; Raquel Tibol, Ed.; Fondo de Cultura Económica: Mexico City, Mexico, 1996; p 507.
11. Gutiérrez, J.L. *From Fresco to Plastics*; National Gallery of Canada: Ottawa, Canada, 1956; pp 56-8.
12. Lodge, R. In *Pre-prints of Papers at the Sixteenth Annual Meeting*; The American Institute for Conservation for Historic and Artistic Works: New Orleans, LA, 1988; p 120.
13. Brown, K. and Crawford, F. U.S. patent, 1,234,921, 1928.
14. Brooks, L. *Oil Painting...Traditional and New*; Reinhold Company Corporation: New York, 1959; p 133.
15. Barclay, M.H. In *The Crisis of Abstraction in Canada: The 1950s*; L. Muir, P. Morriset, Eds.; National Gallery of Canada: Ottawa, Canada 1992; pp 205-231.
16. Gutierrez, J.L. *From Fresco to Plastics*; National Gallery of Canada: Ottawa, Canada, 1956; p 57.
17. Siqueiros, D.A. *Cómo se pinta un mural*; Ediciones Taller Siqueiros Venus y Sol: Cuernavaca, Mexico, 1977; p 146.
18. Jansen, L. *Synthetic Painting Media*; Prentice-Hall: Englewood Cliffs, NJ, 1967; pp 98-106.
19. Marontate, J. Ph.D. thesis, Université de Montréal, Montréal, QC, 1996; p 65
20. Siqueiros, D. A. *Cómo se pinta un mural*; Ediciones Taller Siqueiros Venus y Sol, 1977: Cuernavaca, Mexico, 1977, p 146.
21. Selwitz, C. *Cellulose Nitrate in Conservation*; The Getty Conservation Institute: Los Angeles, CA, 1988; p 23.
22. Horie, C. *Materials for Conservation*; Butterworth Heinman: London, 1996; p 132.

23. Selwitz, C. *Cellulose Nitrate in Conservation*; The Getty Conservation Institute: Los Angeles, CA, 1988; pp 23, 31.
24. Stewart, R., Littlejon, D., Pethrick, R.A., and Tennent, N.H. In *ICOM Committee for Conservation 11th Triennial Meeting in Edinburg, Scotland, 1-6 September: Preprints*; Janet Bridgland, Ed.; James & James (Science Publishers) Ltd.: London, 1996, pp 967-970.
25. Derrick, M., Stulik, D., and Ordonez, E. In *Saving the Twentieth Century: The Conservation of Modern Materials*; David Grattan, Ed.; Canadian Conservation Institute: Ottawa, Canada, 1993, p 169-182.
26. Snodgrass, A. and Price, B. *The Gettens Collection of Aged Materials of the Artists: The FTIR spectra Library and Catalogue*; Philadelphia Museum of Art, Victoria and Albert Museum, and Witerthur Museum: Philadelphia, PA, 1993.
27. Learner, T. In *Resins Ancient and Modern: Preprints of the SSCR*; The Scottish Society for Conservation and Restoration: Edinburgh, Scotland, 1995; pp 76-84.
28. Cruz Arvea, R. Historian at Siqueiros Hall of Public Art, Mexico City, *unpublished observations*.
29. Soumaya Museum Archives, Documentation on Siqueiros's Artwork; Soumaya Museum: Mexico City, D.F.
30. Siqueiros, D.A. *Retrato de una década: 1930-40*; Museo Nacional de Arte and Instituto Nacional de Bellas Artes: Mexico City, Mexico, 1996; pp 116,131.
31. Mateos, F., Conservator at the National Center of Restoration of Cultural Heritage, National Institute of Anthropology and History, Mexico City, *unpublished observations*.
32. Cuevas, R. Sub-Director of National Center of Conservation and Registration of Artistic Heritage, National Institute of Fine Arts, Mexico City, *unpublished observations*.
33. Micheli, M. *Siqueiros*; Harry N. Abrams, Inc.: New York, 1968; pl. "Portrait of a Dead Child".

Chapter 15

Laser Surface Profilometry: A Novel Technique for the Examination of Polymer Surfaces

S. A. Fairbrass and D. R. Williams[1]

Particle Technology, Department of Chemical Engineering, Imperial College, London SW7, United Kingdom

Laser Surface Profilometry is a non-destructive technique that requires little or no sample preparation and takes place in ambient laboratory conditions. Samples of poly(vinyl chloride) were artificially aged by exposing them to high levels of ultra violet light. A near infrared laser was used to produce topographical images of the changes in surface morphology, both on ageing and after simple conservation cleaning techniques. Cumulative percentage graphs of R_a data were shown to be a sensitive indication of any changes, which had taken place. The percentage of reflected laser light was also collected and maps of surface chemical changes were produced. This technique has proved to be a valuable tool for the examination of museum objects.

There are many, specific problems that arise out of the need to examine museum objects. It is often the case that the object itself is unique or that very little of the original remains intact. In this case, the opportunity to take samples from the actual artifact or to use any technique that is destructive may be forbidden. This prohibition includes such ubiquitous examination methods such as scanning electron microscopy where the sample may have to be gold or carbon coated and submitted to a high vacuum. The use of a laser, which provides an image of the surface without the need for sample preparation, has provided fresh scope for the close examination of

[1]Corresponding author.

surfaces. In addition to this, because the equipment allows for an item to be treated in situ, repositioning is no longer a problem and it is now possible to examine and evaluate the results of conservation treatments.

Experimental

Samples of plasticised poly(vinyl chloride) were exposed to high levels of light in order to age the surfaces artificially. Plasticised PVC was chosen for this study as it represents on of the most commonly used polymers and is also known to change and deteriorate noticeably over time. [1,2] Light ageing took place in London, where a bank of lights consisting of a mixture of florescent tubes and ultra-violet lights were used, and also in Los Angeles where the samples were simply left outdoors in bright sunlight, high humidity and significant atmospheric pollution. The Blue Wool Scale, a simple light dosimeter produced by the textile industry to monitor dye fading [3] was used to give some indication of the light exposure in both locations. This scale ranges from 1 to 8, with wool sample #1 being the most light fugitive and sample #8 being the least. The PVC was exposed to enough natural or artificial light to fade Blue Wool Sample #4, 5 or 6. After aging, the PVC samples were imaged using Laser Surface Profilometry. The top, exposed surfaces of the aged samples were also cleaned by using either a 2% non-ionic surfactant solution or de-ionized water. Surface images were made of the samples before and after cleaning.

The Profilometer

The profilometer consisted of an infrared laser (780nm/0.2 microwatts) and three micro stages, *x,y and z*, that could be controlled to within an accuracy of 0.1μm. The auto focusing laser was attached to the *z-axis* and focused onto the top surface of a sample of PVC.

The laser worked within three ranges, ± 400μm which could be used to produce an image of the surface of fine glass paper, ± 40μm which would be the range for writing paper and ± 40μm which is the range for most synthetic polymeric materials. It has a best resolution of 5nm in the *z-axis* over a range of ± 4μm. The stand off distance of the laser from the sample was 10mm. The samples were held on a flat platform, normal to the laser light, and capable of being moved in the *x and y* directions. The stages were moved so that the laser light, <1μm in diameter progressed in a raster scan across the surface of the sample in 1μm steps. Typical sample areas measured 250 x 250 μm.

At each step of the scan the light reflected from the surface was recorded through a collimating lens within the laser module. The height or depth of any asperities on the surface of the PVC were calculated from the displacement of the lens when the light was at a maximum. Gross errors of form, such as sample curvature were removed by filtering each scanned line in the y-direction using Chebyshev polynomials. [4] The resulting data matrix was translated into a visual topographic image using Spyglass Transform (Adept Scientific Micros Systems).The percentage of the laser light reflected back from the surface at each step was also recorded. This was displayed as a map of surface reflectivity at 780nm.

The Centre Line Average

The Center Line Average or R_a of each line in the y-direction was calculated. This parameter is one of the most commonly used measurements to describe surface roughness. If a line xy of length L is drawn across a profile then the sums of the areas above and below the line can be calculated. Ra is the distance of a second line, parallel to the line xy, where sums of the areas above and below the line are equal.

$$R_a = 1/L \int_0^L |y(x)|\, dx$$

where L = sample length
    y = the height of the surface above a mean line at distance x from the origin.

Results and Discussion

Surface Roughness

Figure 1 shows the surface of an unaged piece of poly(vinyl chloride). All of the details on the surface are between ± I micron in height or depth. The surface has several scratches and gouges from dust and handling but is otherwise reasonably smooth. Figure 2 shows a similar sample, which had been exposed to bright sunlight and pollution in Los Angeles until Blue Wool Standard #6 had faded.

The surface of the polymer is now quite rough and pitted. The series of holes approximately 20μm in diameter, which can be seen near the centre of the image are one of the characteristic features that are found when plasticised PVC is aged. A layer of sub-micron cavitation occurs in randomly spaced centres about 100Å below the surface of most rigid polymers during ageing[5]. Because of the softer viscoelastic nature of plasticised polymers an increase in the cavitation as the ageing process progresses results in holes appearing on the surface. It should be noted that most museum surveys document changes in the appearance of plastics on ageing by recording a definite yellowing of dulling of the surface. The surface changes shown

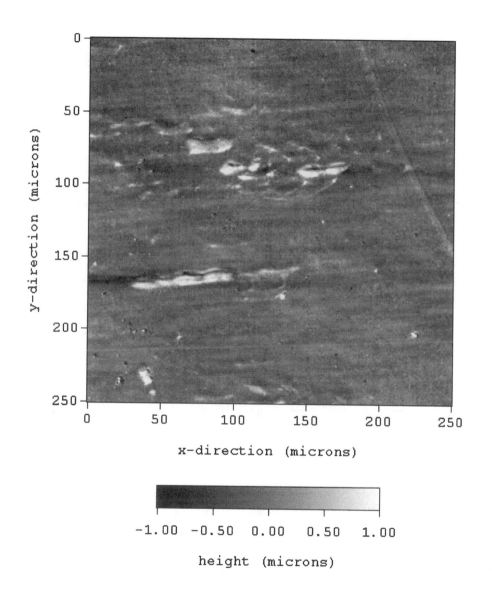

Figure 1 Topographic image of the surface of unaged PVC

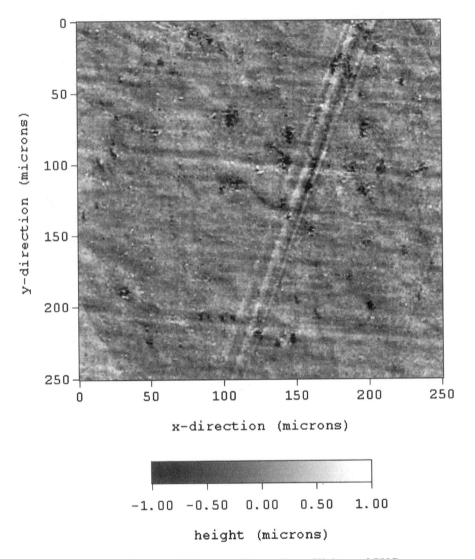

Figure 2 Topographic image of the surface of light aged PVC

here have occurred in PVC films that showed no visual sings of colour change and were still quite flexible.

The cumulative percentage graph of R_a data form the new PVC and samples aged under artificial light in London is shown in Figure 3. Average surface roughness can be estimated from the x-axis value at 68% (± one standard deviation about the mean). Unaged PVC has an average surface roughness value of below 0.12μm. Exposure of the samples to light causes in increase in the surface roughness of between 0.06 and 0.07μm, with the roughness increasing as a direct consequence of light dosage.

Conservation Treatments

Figure 4 shows the surface of a sample of PVC that had been left outside in the sun in Los Angeles until Blue Wool Standard #5 had faded. This sample is just starting to show the general surface roughness, which was so apparent in Figure 2. There is also a large hole, approximately 50μm in diameter towards the right of the centre. This has been caused either by a particle of grit or, as is more likely, the juncture of several small, sub-surface holes.

The exposed surface of the sample was gently cleaned with a cotton wool swab dipped in a non-ionic surfactant solution. Figure 5 shows the same surface after treatment. The first thing to notice is the accuracy of re-positioning. This is because the equipment allows for the laser to be raised up on the z-stage, which the sample is left in position of the x,y-stages. So long as the sample is not moved during treatment the laser can then be re-focused on exactly the same spot and a second image of the surface can be recorded.

The same minor asperities are still present in Figure 5 and several of the bumps and ridges can be traced on both images. The pit near the centre has disappeared, however. There are several explanations for this phenomena and it is thought that the action of the surfactant on a viscoelastic material has caused the release of stress and the recovery of the original surface features. The dirt particles and small pits in the top left corner have also been removed.

Similar surface changes, which could be regarded as beneficial, were observed on all of the aged samples that were cleaned with a surfactant solution. Comparison of Figures 4 and 5 however shows that only gross deformations such as holes and pits were removed. From this it can be concluded that cleaning the surface of a soft, plasticised polymer with a mild surfactant solution can be advantageous and may help prolong the life of a museum object.

Conclusions

The use of a laser to provide topographical images of the surface of polymeric materials has opened up new possibilities for the safe examination of museum objects. Because this is a non-contact technique, using a low power laser, there is no danger of damage occurring to the surface. The equipment also allows for objects to be treated in situ with total accuracy in repositioning. The final images provide

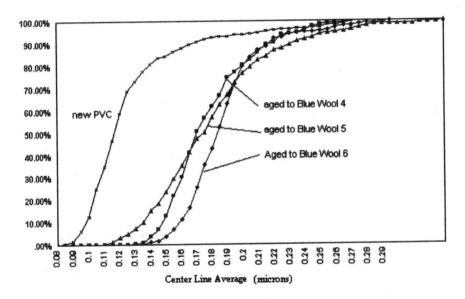

Figure 3 Cumulative percentage graph of the Ra data from four PVC samples

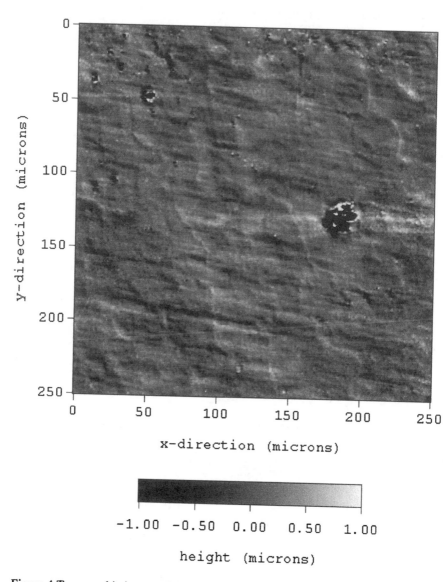

Figure 4 Topographic image of the surface of PVC aged to Blue Wool Standard #5

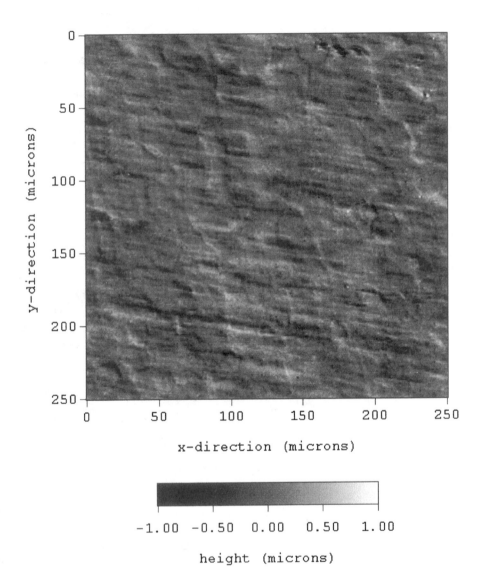

Figure 5 Topographic image of Figure 4 after cleaning with a 2% non-ionic surfactant solution.

immediate, detailed visual information about the condition of an object and also the effects of conservation treatment. The cumulative percentage graph of R_a data supplies quantitative information and confirms subtle changes which may not be immediately apparent.

Acknowledgements

Sheila Fairbrass would like to thank her doctoral supervisor Daryl Williams, Professor Brain Briscoe and the members of the Particle Technology Group, Chemical Engineering Department, Imperial College, UK, the Science Museum London for their financial support and Teri Schaeffer from the Los Angeles Museum of Art, CA for helping to age the polymer samples.

References

1. *Principles of Polymer Systems 3rd Ed*; Rodriguez, F.; Hemisphere Publishing Company: 1989
2. Shashua, Y. *A passive approach to the conservation of poly(vinyl chloride)*; ICOM-CC 11th Triennial: 1996, Vol.II
3. (ISO)R 105 Blue Wool Fading Standard
4. *Handbook of Surface Metrology*; Whitehouse, D.J.; IOP Publishers Ltd:1994
5. *Surface Roughness*; Thomas, T.R.; Longman Inc.;1982

Author Index

Subject Index

A

Accelerated aging
  analyses of soluble degradation products of cellulose to compare effects of aging conditions on paper, 25–26
  changes in percent elongation of vulcanized rubber as function of QUV exposure, 183*f*
  comparing aging conditions by changes they induce, 25
  definition, 24
  illustration of apparatus for cellulose acetate, 148*f*
  method for vulcanized natural rubber, 168
  random cracking of rubber samples matching typical cracking of museum artifacts, 184
  study of cellulose acetate, 147
  vulcanized natural rubber samples, 180, 181*f*, 182*f*
  See also Naturally and accelerated aged cellulose; Vulcanized natural rubber artifacts
Acid, degradation of cotton and flax, 12–13
Aged cellulosic textiles
  Arrhenius model to estimate service life and failure distribution, 16–17
  Shroud of Turin, 15–16
  See also Historic materials from cellulosic fibers
Aging
  accelerated studies of cellulosic fabric, 19
  complicating polymer identification, 130
  properties of pre-Columbian versus contemporary cotton, 16
  simulating to estimate lifetime, 6–7
  See also Historic materials from cellulosic fibers; Naturally and accelerated aged cellulose; Wool, aging

Aging temperature, natural, calculating, 28–29
Air pollution
  nitrogen oxides, 19
  ozone, 18–19
  sulfur dioxide, 18
Alkaline treatment
  ballooning behavior after treatment with NaOH and CS_2, 59, 60*f*
  FT–IR spectra of cotton fibers treated with 18% NaOH, 59, 63, 66
  mordant of dyed cotton fibers, 71
  swelling behavior of cotton fibers, 58–59
  See also Microstructure of historic fibers through microchemical reaction
Aluminum salts as mordants. *See* Cotton fabrics dyed with natural dyes and mordants; Silk fabrics dyed with natural dyes and mordants
Archives, preservation of collections, 23–24
Arrhenius equation, ratio of rates of reaction at two different temperatures, 28–29
Arrhenius model, estimating service life and failure distribution, 16–17
Artifacts. *See* Cellulose acetate artifacts; Vulcanized natural rubber artifacts
Attenuated total reflectance–microscopy–Fourier transform infrared spectroscopy (ATR–microscopy–FTIR)
  ATR–microscopy–FTIR of unexposed, artificially aged, and museum rubber sample, 171*f*, 173*f*
  limited usefulness, 170, 172
  method, 168–169
  minor ability to evaluate vulcanized rubber samples, 180, 184
  See also Vulcanized natural rubber artifacts
Authenticity, preserving textiles, 2–3

Bestsellers from ACS Books

The ACS Style Guide: A Manual for Authors and Editors (2nd Edition)
Edited by Janet S. Dodd
470 pp; clothbound ISBN 0–8412–3461–2; paperback ISBN 0–8412–3462–0

Writing the Laboratory Notebook
By Howard M. Kanare
145 pp; clothbound ISBN 0–8412–0906–5; paperback ISBN 0–8412–0933–2

Career Transitions for Chemists
By Dorothy P. Rodmann, Donald D. Bly, Frederick H. Owens, and Anne-Claire Anderson
240 pp; clothbound ISBN 0–8412–3052–8; paperback ISBN 0–8412–3038–2

Chemical Activities (student and teacher editions)
By Christie L. Borgford and Lee R. Summerlin
330 pp; spiralbound ISBN 0–8412–1417–4; teacher edition, ISBN 0–8412–1416–6

Chemical Demonstrations: A Sourcebook for Teachers, Volumes 1 and 2, Second Edition
Volume 1 by Lee R. Summerlin and James L. Ealy, Jr.
198 pp; spiralbound ISBN 0–8412–1481–6
Volume 2 by Lee R. Summerlin, Christie L. Borgford, and Julie B. Ealy
234 pp; spiralbound ISBN 0–8412–1535–9

The Internet: A Guide for Chemists
Edited by Steven M. Bachrach
360 pp; clothbound ISBN 0–8412–3223–7; paperback ISBN 0–8412–3224–5

Laboratory Waste Management: A Guidebook
ACS Task Force on Laboratory Waste Management
250 pp; clothbound ISBN 0–8412–2735–7; paperback ISBN 0–8412–2849–3

Reagent Chemicals, Ninth Edition
768 pp; clothbound ISBN 0–8412–3671–2

Good Laboratory Practice Standards: Applications for Field and Laboratory Studies
Edited by Willa Y. Garner, Maureen S. Barge, and James P. Ussary
571 pp; clothbound ISBN 0–8412–2192–8

For further information contact:
Order Department
Oxford University Press
2001 Evans Road
Cary, NC 27513
Phone: 1-800-445-9714 or 919-677-0977

Highlights from ACS Books